D1514737

changing the way the world learns

To get extra value from this book for no additional cost, go to:

http://www.thomson.com/wadsworth.html

thomson.com is the World Wide Web site for Wadsworth/ITP and is your direct source to dozens of on-line resources. *thomson.com* helps you find out about supplements, experiment with demonstration software, search for a job, and send e-mail to many of our authors. You can even preview new publications and exciting new technologies.

thomson.com: *It's where you'll find us in the future.*

Contemporary Readings for General Biology

Fifth Edition

To Accompany Starr and Taggart's
BIOLOGY
The Unity and Diversity of Life

and

Starr's
BIOLOGY
Concepts and Applications

John W. Crane
Washington State University

Wadsworth Publishing Company
I(T)P® An International Thomson Publishing Company

Belmont, CA • Albany, NY • Bonn • Boston • Cincinnati • Detroit •Johannesburg • London • Madrid
Melbourne • Mexico City • New York • Paris • San Francisco • Singapore • Tokyo • Toronto • Washington

Biology Editor: Jack Carey
Assistant Editor: Kristin Milotich
Production Editor: Jennie Redwitz
Print Buyer: Stacey Weinberger
Permissions Editor: Peggy Meehan
Compositor: Jeffrey Sargent, Pacific Publications
Printer: Malloy Lithographing, Inc.
Cover Photos: © Art Wolfe/Tony Stone Images (left); © Thomas D. Mangelsen (right)

Printed in the United States of America
  2 3 4 5 6 7 8 9 10

For more information, contact Wadsworth Publishing Company, 10 Davis Drive, Belmont, CA 94002,
or electronically at http://www.thomson.com/wadsworth.html

International Thomson Publishing Europe
Berkshire House 168-173
High Holborn
London, WC1V 7AA, England

International Thomson Editores
Campos Eliseos 385, Piso 7
Col. Polanco
11560 México D.F. México

Thomas Nelson Australia
102 Dodds Street
South Melbourne 3205
Victoria, Australia

International Thomson Publishing Asia
221 Henderson Road
#05-10 Henderson Building
Singapore 0315

Nelson Canada
1120 Birchmount Road
Scarborough, Ontario
Canada M1K 5G4

International Thomson Publishing Japan
Hirakawacho Kyowa Building, 3F
2-2-1 Hirakawacho
Chiyoda-ku, Tokyo 102, Japan

International Thomson Publishing GmbH
Königswinterer Strasse 418
53227 Bonn, Germany

International Thomson Publishing Southern Africa
Building 18, Constantia Park
240 Old Pretoria Road
Halfway House, 1685 South Africa

ISBN 0-534-51632-7

CONTENTS

I
THE CELLULAR BASIS OF LIFE   1

II
PRINCIPLES OF INHERITANCE   17

III
PRINCIPLES OF EVOLUTION   39

BEHAVIOR

PREFACE

For the memories of Nancy Antonia Crane, who always listened to the owl

The quality of all life depends on a human understanding of the intricacies of the biotic/abiotic connection, for we not only control this connection, but also depend upon it for our own existence. Biologists now recognize the importance of diversity, both biotic and abiotic, in the maintenance of properly functioning and viable ecosystems. But the owl and salmon speak—and we still refuse to listen! Their diminishing populations are trying to inform us that if we continue to reduce the viability of their ecosystems through habitat destruction and pollution, we reduce our own fitness as well. Based on an objective assessment of our environmental quality, the outlook appears dim. Earth, air, and water qualities continue to suffer and are in worse shape than they have ever been—and we still don't pay attention. Listen to the owl—the future quality of all life depends upon the message!

New information regarding other biological concerns also makes headlines. News emanating from the biological sciences affects our lives on a scale never before imagined. Almost daily we hear of diseases such as AIDS, lupis, and hemorrhagic fever, for which no cures exist. AIDS, which results from destruction of the human immune system by the HIV virus, results in almost certain death. Research into the causes of cancer, lupis, heart disease, and a host of other diseases continues. Although surgical, drug, and other forms of therapy aid in treatment, prevention and finite cures still elude us. Lung cancer and heart diseases in women are increasing at alarming rates and will cause more deaths in the United States in 1996 than at any time in our history. And yet there is hope. Investigations in cellular biology, biochemistry, genetic engineering, and physiology are producing new medical breakthroughs. Genetic engineering, encompassing the fields of gene manufacture and gene substitution, holds much promise for preventing, and perhaps curing, various diseases. In fact, genetic treatment of human patients has already begun.

As we explore new biological constructs, we also continue to examine older concepts such as evolution in the attempt to refine these doctrines in light of modern scientific inquiry. Efforts to explain the absence of observable continuity between species have resulted in the concept of "punctuated equilibrium"—evolution by bursts rather than by gradual change. The reexamination of Darwin's hypotheses reveals that, with some modifications, his doctrines hold true.

In producing the Fifth Edition of *Contemporary Readings for General Biology*, we retain our original goal: to demonstrate the importance, dynamic nature, and impact of the biological sciences for all of us—including the owl.

John W. Crane

CROSS-REFERENCE GUIDE

to Starr and Taggart's Biology: The Unity and Diversity of Life, *Seventh Edition*

This cross-reference guide correlates selected readings in this book with units and chapters in Starr and Taggart's *Biology: The Unity and Diversity of Life,* Seventh Edition.

CROSS-REFERENCE GUIDE

to Starr's Biology: Concepts and Applications, *Third Edition*

This cross-reference guide correlates selected readings in this book with units and chapters in Starr's *Biology: Concepts and Applications,* Third Edition.

I

The Cellular Basis of Life

Human life, as does most, begins with but a single cell. In our case we each begin as a cell containing 46 chromosomes—23 from each parent. From this beginning we eventually mature to an individual containing an almost uncountable number of cells, many of which are far different from the original. The eventual fate of each cell is, of course, death; perhaps produced, as some biologists believe, by an innate genetic program. Apoptosis, or programmed cell death, is currently receiving scientific attention regarding the molecular mechanisms causing the event. For not only does it apparently cause cellular death, apoptosis is believed to be responsible for preventing webbed fingers in humans and may underlie metamorphosis of amphibians and insects. Apoptosis may also have a "dark side" as recent findings implicate it as a potential cause of cancer and a possible factor in AIDS and other autoimmune diseases. Why and how these cells evolve from a single ancestor to widely divergent cellular components and then perish is receiving much scientific investigation as biologists begin to probe the genetics of programmed cell death.

Relative to apoptosis are many unanswered questions regarding human fetal development. The question of how, in nine months, a zygote measuring about 0.025 of an inch develops into an infant containing a myriad of cells is receiving much attention. Biologists are currently focusing on "master control" genes that may program other cellular systems to change directions in their eventual fates. In fact, certain "master control" genes are implicated in producing morphological variations when inserted into developing embryos. As developmental research continues we will begin to understand how all living systems integrate the various genetic messages that eventually culminate in the production of an adult form.

Also, as population growth continues, we are observing a concomitant increase in the number of older members within our communities. As the population ages, medical attendants are seeing a rise in the number of patients with Alzheimer's and other age-related diseases. Very recent medical/genetic research suggests that at least one form of Alzheimer's disease can be traced to a genetic cause. Other studies suggest that Alzheimer's disease may result when proteins produced by cells and known as "complements" are unleashed against the fragile cells of the brain itself.

But all is not gloom. Potential treatments are on the horizon for many cellular diseases, running the gamut from AIDS and lupus to sickle-cell anemia. Of course much remains to be done, but the scientific community continues to gain momentum in its attempt to understand the basic unit of life—the cell.

A Time to Live, a Time to Die

Biologists probe the genetics of programmed cell death

By Carol Ezzell

"First, you murder," Michael O. Hengartner forthrightly told a horde of expectant faces. "Next, you get rid of the body. Then, you hide the evidence," he explained, pacing back and forth in the dimly lit room.

Hengartner wasn't instructing a group of apprentice hit men. Instead, the Massachusetts Institute of Technology (MIT) biologist was addressing a gathering of cancer researchers, detailing the functions of a recently identified set of genes that controls life's only inevitable process: death.

Together, Hengartner and his MIT colleagues constitute one of scores of research teams around the world who are reviving scientific interest in the molecular mechanisms of a phenomenon called apoptosis, or programmed cell death. Among other things, this phenomenon (pronounced apa-tosis, with the second "p" silent) prevents humans from having webbed fingers and eliminates cells of the immune system that can't tell "self" from "nonself." It also underlies metamorphosis—the magic wand that turns caterpillars into butterflies and tadpoles into frogs. In adults, it phases out old body cells so they can be replaced by new ones.

Over the past year, biologists from a range of disciplines have uncovered evidence that this seemingly salutary process has a dark side. Several new studies suggest that apoptosis can play roles in AIDS and autoimmune diseases; others indicate that disruptions in the usual orderly progression of apoptosis lead to the uncontrolled cell growth of cancer.

Apoptosis—which means "dropping off" in Greek—was first described in 1951 as a step in animal development. The process takes its name from its appearance as it unfolds under the microscope: Within minutes, cells undergoing apoptosis shrink and shed tiny, membranous blebs that neighboring cells quickly gobble up, mirroring Hengartner's colorful description. In contrast, during necrosis—cell death arising from injury—cells swell for hours and then burst, spraying their contents about as a chemical signal that attracts immune-system cells to fight the injurious microbe or substance.

In the late 1960s and early 1970s, researchers began gathering evidence that apoptosis occurs as part of the normal turnover and replacement of worn-out tissues in adult organisms. They discovered that apoptosis resem-

> They discovered that apoptosis resembles suicide in some ways: Old cells actively participate in their own demise by turning on genes and making new proteins that will shortly cause their death.

bles suicide in some ways: Old cells actively participate in their own demise by turning on genes and making new proteins that will shortly cause their death.

Since the mid-1980s, cell biologists and geneticists have started sorting out the causes and implications of apoptosis in a wide range of animals, including humans. Last spring they began reporting evidence of the role played in apoptosis by the cancer-causing c-myc gene—named for its initial discovery in myelocytomas, tumors consisting of tightly packed bone marrow cells.

The c-myc oncogene becomes overactive in a wide range of mammalian tumors, including human cancers of the breast, bladder, colon, lung, and cervix (SN: 6/1/91, p.347). In many cases, c-myc's hyperactivity begins when a cell inexplicably creates extra copies of the gene, reproducing it over and over within the cell nucleus.

Because cells with such c-myc amplifications grow and divide nonstop—and further, because the c-myc gene encodes protein-containing regions that can bind to DNA—scientists hypothesize that c-myc regulates other genes involved in cell division. Ironically, Gerard I. Evan of the Imperial Cancer Research Fund Laboratories in London and his colleagues reported in the April 3 CELL that c-myc can also cause apoptosis under certain conditions .

Evan's group found that while laboratory-cultured cells with hyperactive c-myc genes can grow faster than cells with less active c-myc genes, they also die faster than those cells when deprived of growth medium. Moreover, the researchers noted, the cells with overactive c-myc genes died with all the visible hallmarks of apoptosis.

Evan and his co-workers conclude that c-myc functions as a two-edged sword: While it usually acts to keep a healthy cell dividing, it can also trigger cell death if outside conditions aren't right for continued cell proliferation or if the cell has become genetically damaged. In this way, c-myc can function as a built-in cellular self-destruct mechanism.

So how does c-myc cause cancer? According to a model developed by Evan and his co-workers, damage to the c-myc gene—caused either by slips in the DNA-repair machinery or by environmental injury—usually results in cell death. But some cells sustain such genetic damage and go on to develop a mutation that activates, or turns on, a second gene. This second gene somehow overrides c-myc's death command, allowing the cells to grow into tumors.

Two papers in the Oct. 8 NATURE provide evidence that this second gene is bcl-2, an oncogene named for its initial discovery in human immune-system cancers called B-cell lymphomas. In the first paper, a team led by Douglas R. Green of the La Jolla (Calif.) Institute for Allergy and Immunology reports that death-prone cells containing extra c-myc genes survive much longer following insertion of an activated bcl-2 gene, which produces a protein with unknown function.

"In the absence of bcl-2, c-myc induces death," summarizes Green, "but in the presence of bcl-2, there's no death." Cancer results, he asserts, "not just because the [mutated] cells grow faster, but also because they die more slowly."

In the second paper, a team led by the Imperial Cancer Research Fund's Evan reports similar results and provides evidence suggesting that the bcl-2 mutation can help cancer cells resist the deadly effects of chemotherapeutic drugs. Many such drugs kill cancer cells by causing them to undergo apoptosis.

Evan's group administered the anticancer drug etoposide, also known as VP16, to death-prone rat cells genetically engineered to contain the activated bcl-2 gene. The researchers found that the bcl-2 gene prevented many of the cells from undergoing apoptosis and delayed its onset in others.

Further evidence that bcl-2 increases the resistance of cancer cells to chemotherapy is published in the Oct. 1 CANCER RESEARCH. Toshiyuki Miyashita and John C. Reed of the University of Pennsylvania School of Medicine in Philadelphia inserted copies of the activated human bcl-2 gene into mouse lymphoid tumor cells. They found that the genetically engineered cells survived a dose of the steroid drug dexamethasone roughly 100 times larger than that required to kill cells lacking the bcl-2 gene. Moreover, the cells resisted death induced by several other chemotherapeutic drugs, including the widely used cancer therapies vincristine and methotrexate.

The findings "may open the door to a whole new approach for the treatment of cancer," says Reed, who is now at the La Jolla (Calif.) Cancer Research Foundation. "If you could use drugs to reduce the expression of bcl-2 [in cancer cells], you might make the cells more sensitive to existing chemotherapeutic drugs," he suggests.

Reed and his colleagues are working with Genta, Inc., a San Diego-based biotechnology company, to develop so-called antisense drugs to block the activity of bcl-2. Antisense drugs—which consist of the same chemical building blocks that make up the genetic material DNA—turn off specific genes by binding to and inactivating messenger RNA, the intermediate compound that genes use to tell a cell to make a given protein (SN: 2/16/91, p.108).

Reed says initial tests in laboratory-cultured cells show that antisense drugs that target bcl-2 make cancer cells more vulnerable to apoptosis induced by chemotherapeutic drugs. "We're hoping to get our [bcl-2] antisense drug into clinical trials soon," says Reed. "We'd love to see if we could get it to work [in cancer patients]."

"There's a possibility that in all [the processes that turn cells cancerous] there may be mechanisms that favor cell death," adds Green. "If other genetic changes override that, you get full-scale transformation [into a cancer cell]."

In the meantime, MIT's Hengartner has found that the tiny roundworm *Caenorhabditis elegans* has a gene that resembles human bcl-2. He reported last month that the structure of bcl-2 is similar to that of a roundworm gene called ced-9, for *C. elegans* death (SN: 10/10/92, p.229). Moreover, like bcl-2, ced-9 protects cells from programmed cell death, Hengartner and his colleagues reported in the April 9 NATURE.

Hengartner says that ced-9 regulates the activity of two other genes, ced-3 and ced-4, that actually cause cells to undergo apoptosis. When ced-9 is "on," it shuts off ced-3 and ced-4, allowing a cell to live. But when ced-9 is inactivated by a mutation, ced-3 and ced-4 start up, prompting a cell to commit suicide.

This feedback mechanism ensures that so-called stem cells in a developing roundworm die when they are no longer needed, says Hengartner. Scientists know that the minuscule roundworm generates 1,090 cells during its embryonic development. However, 131 of these cells die, so an adult roundworm consists of exactly 959 cells.

Hengartner's team has shown that roundworms with an abnormally activated ced-9 gene develop superfluous body parts, presumably because the extra 131 cells never die. In contrast, the researchers report, the offspring of roundworms lacking functional ced-9 genes die as embryos, evidently because the ced-3 and ced-4 genes functioned unchecked, killing all of the young organism's cells prematurely.

"Ced-9 is the switch between life and death" in the developing roundworm, concludes Hengartner.

Two studies published earlier this year demonstrate that the mammalian immune system may employ a similar set of cell-death genes. In the first study, a group led by Shigekazu Nagata of the Osaka Bioscience Institute in Osaka, Japan, has found that mice genetically predisposed to an affliction resembling the human autoimmune disease systemic lupus erythematosus (SLE) have defects in a protein required for apoptosis in white blood cells.

Accordingly, Nagata and his colleagues report in the Mar. 26 NATURE, the mice fail to purge themselves during embryonic development of white blood cells that attack their own tissues. As a result, the animals develop the swollen lymph glands, lethargy, and tissue damage characteristic of lupus.

The results reported by Nagata's team "are the first hint of a cell-death link with a real disease model," comments Green. "It looks like a gene that is involved in the programmed cell death process is defective in this strain of mouse with horrendous autoimmune problems."

In the second paper, which appeared in the July 10 SCIENCE, a group led by Frank Miedema of the University of Amsterdam in the Netherlands reports evidence that AIDS resembles the other side of the same coin. Miedema and his colleagues took white blood cells called T-cells from male AIDS patients. When they stimulated the cells' CD3 receptors using antibodies, up to one-fourth of the cells committed suicide by apoptosis. In contrast, the antibody treatment failed to induce significant levels of apoptosis in T-cells isolated from men not infected with the AIDS-causing HIV virus.

Miedema and his colleagues suggest that HIV infection "hyperactivates" T-cells, giving them a hair-trigger tendency toward suicide. They say this mechanism may explain why AIDS patients show a decrease in all types of T-cells, not just those bearing the CD4 receptor that HIV uses to enter and infect some T-cells.

Developments such as these signal renewed interest in the study of cell death among researchers from a variety of fields, say many biologists. "This will be a very fruitful area of research for some time to come," predicts Reed.

THOUGHT EVOKERS

- Explain the concept of apoptosis.

- Why has it been suggested that apoptosis might have a "dark side" regarding the human autoimmune system?

- Describe how the c-myc oncogene is believed to be involved in causing cancer.

Body building from scratch

Breakthroughs in developmental biology teach us how we grow

By Traci Watson

It is one of life's most splendid mysteries: In nine months, a 0.025-inch fertilized egg somehow turns into a kicking, crying human baby. No topic could be more intriguing—or less understood. Humans start life small and inglorious, as "little, sluglike things," says one developmental biologist. But the "slug" somehow conjures up eyes, a brain, and other organs. How it does so is finally being worked out by scientists.

Thanks to new laboratory methods, scientists are gaining awesome insights into what makes the embryo tick. Study of the embryo "is enormously fun because we're learning so much so fast," says Cliff Tabin, a developmental biologist at Harvard University. In the past six months alone, researchers have done important work on genes for head and eye formation. Last week, scientists reported the first discovery of genes directing the left-right orientation of

> **The resulting mouse pups had normal bodies below the neck. Above the neck, they had . . . nothing.**

the body. And in the ultimate sign of trendiness, biotech companies have recently taken an interest in developmental biology, once considered a backwater of useless knowledge.

Much of the attention is being lavished on "master control" genes that tell parts of the body how to form. In one startling finding, a Swiss research team reported in March that it had created fruit flies with eyes in the oddest places—on their wings, their legs, their antennae. To make these all-see-ing flies, the scientists simply gave a fly gene called eyeless—probably a master-control gene—to fly embryos. All by itself, eyeless tells a growing fly to make an extra eye. Mice and humans also have a version of eyeless.

Growing heads. Dramatic as eyeless is, it seems mundane next to the gene named Lim-1. In a paper published in March, researchers at the M. D. Anderson Cancer Center in Texas reported that they had created very young mouse embryos lacking Lim-1. The resulting mouse pups had normal bodies below the neck. Above the neck, they had . . . nothing.

Examining these freakish results, the M. D. Anderson scientists realized that they had found an important clue to the powerful, mysterious region called the head organizer. Located in the very young embryo, the organizer somehow tells a group of cells, "Become a head," which they do. The new study shows that one of the organizer's key agents is Lim-1, a finding that will help reveal all the steps required to grow a head.

Just because an embryo has a head, however, doesn't mean it knows how to proceed. A developing animal must also figure out which way is up and which down, which way is back and which front. And the genes that control those commands are now being tracked down. Chief among them are the 39 "Hox" genes. So fundamental is this genetic family that it is found in nearly the exact same form in all animals with backbones. The list of body parts the Hox genes help form reads like the index of an anatomy textbook: nervous system, sexual organs, skeleton.

A barrage of recent experiments have shown that at least some of the Hox genes help the embryo tell head from tail. Certain Hox genes, it seems, switch on only at the embryo's front; others switch on at the rear. By looking at which Hox genes are on, cells know what to become. A series of recent studies show that if a single Hox gene is removed from one part of the embryo, that part develops like the section closer to the head.

But knowing front from back isn't enough. An embryo also needs to know its left from its right, because the two halves of an animal aren't mirror images. The heart is slightly tilted to one side, for example. This isn't trivial: People born with reversed organs often suffer cardiac ailments and infertility.

Right and left. Last week, the journal *Cell* carried a report that has other scientists cheering: A research team led by Harvard's Tabin has discovered three proteins that control left-right patterning. The researchers found that one of the proteins appears on the right side of the embryo's chest. The other two proteins appear only on the left side. This lopsided pattern prompts the budding heart to tilt leftward and probably affects the slant of other organs, too.

Even more is known about how limbs form. For example: Take a look at your arm. When it was forming, the young cells had to know whether they were closer to the thumb or the pinkie. The signal for that information was a protein now called sonic hedgehog (after a computer game character). A second protein told the arm cells whether they were closer to the shoulder or the fingertips; a third protein, whether they were closer to the back or the palm of the hand. If just one protein had been missing, your arm could have had two palms or no thumb or a hand joined to the shoulder.

This kind of detailed information has not gone unnoticed by other scientists. To evolutionary biologists, the genes and proteins at play in the

---

If just one protein had been missing, your arm could have had two palms or no thumb or a hand joined to the shoulder.

---

embryo are a gold mine: They provide a window on how new types of animals arose, even in the distant past.

When Sean Carroll of the Howard Hughes Medical Institute at the University of Wisconsin at Madison began studying how insect wings develop, he was surprised to find that the butterfly's gorgeous sails and the fruit fly's plain little flappers are sculpted by many of the same genes. Butterfly and fly embryos simply turn on their wing genes in different patterns. So one key to forming new bodies—and thus new species—lies in how embryos play the genes they've already got. It's only fitting that the miracle of life's diversity is being revealed by the miracle of how each creature—human, butterfly, bird—comes to life.

THOUGHT EVOKERS

- What is a "master control" gene?

- Why are the 39 "Hox" genes considered fundamental for animal development?

- Describe what is meant by the phrase "left-right patterning."

Anatomy of Alzheimer's

Do immune proteins help destroy brain cells?

By Kathy A. Fackelmann

"You keep the parts of the complement system separated, like the parts of an atomic bomb," says Patrick L. McGeer of the University of British Columbia in Vancouver. "You begin to assemble the complement proteins and you're likely to do a lot of damage."

McGeer is a neuroscientist with a healthy respect for the complement system—a particularly lethal group of about 25 proteins in the immune system that helps destroy disease-causing microorganisms. He and other scientists now suspect the complement system may play a role in Alzheimer's disease.

During the 1980s, McGeer and other researchers gathered evidence suggesting a link between the complement system and Alzheimer's, a cause of dementia in up to 4 million people in the United States. Now, a California team reports that brain cells taken from people with Alzheimer's disease appear to manufacture certain complement proteins.

These findings hint that Alzheimer's disease results when the explosive complement system is unleashed not on a microbe, but against the fragile cells of the brain itself. The theory remains highly speculative, yet an Arizona team has taken the next step: They are testing an Alzheimer's treatment that may prevent the immune system's misguided attack.

The urgent pace at which Alzheimer's research moves forward may be explained by the fact that as yet no cure exists for the disease.

Dementia refers to a group of symptoms, such as forgetfulness and confusion, that are often associated with old age. Yet many researchers believe that the loss of mental agility is not a normal consequence of aging, but rather the result of a disease process such as Alzheimer's, stroke, Huntington's disease, or Parkinson's disease.

Alzheimer's disease first becomes evident in the mundane tasks of everyday life: A middle-aged woman forgets where she left the house keys. In most cases, people attribute such a lapse to a particularly busy day, an emotional upset, or a variety of other distractions. And in most cases, that explanation proves correct.

For the person with Alzheimer's disease, however, such lapses gradually get worse. The afflicted person may need to compile extensive lists simply to get through a routine day. He or she may not remember how to get home from the grocery store or how to write out a check. Family members may notice a change in behavior or personality. For example, people with Alzheimer's disease may suddenly start to act in a belligerent or agitated manner. As the disease progresses, people with Alzheimer's disease may lose control of their bladder, need assistance with grooming, eating, and other daily activities, and eventually lapse into a vegetative state.

Despite the devastation it creates, scientists still know very little about what causes Alzheimer's disease, which was named after the German physician Alois Alzheimer. Since his detailed description of the disorder in 1907, researchers have put forward a wide range of theories to explain its cause. Some scientists believe a slow virus or environmental poison may initiate the disease. Others have noted that Alzheimer's disease runs in families and believe that some people inherit a predisposition to the illness.

Under magnification, brain tissue taken during autopsy from a victim of Alzheimer's disease shows abnormal yarn-like deposits, called neurofibrillary tangles. In addition, the gray matter of the brain will appear pocked with "plaques." These plaques are made up of dying nerve cell components surrounding a core of beta amyloid protein, a fibrous material that is the subject of an intense, ongoing research effort (SN: 3/7/92, p.152). Both the plaques and the tangles occur in areas of the brain known to be involved in memory and intellectual functioning.

Interest in the complement system was piqued among Alzheimer's investigators during the late 1980s, when Canadian researchers led by McGeer and European teams began to study brain tissue taken during autopsy from people with the disease. The researchers discovered complement proteins buried within the telltale plaque and neurofibrillary tangles. This finding galvanized a team in Los Angeles.

"That evidence led us to wonder if the complement system, which is such a destructive system, could potentially be involved in the neurodegenerative events in the Alzheimer brain," says Steven A. Johnson, a neuroscientist at the University of Southern California (USC) in Los Angeles.

This line of reasoning by Johnson, McGeer, and others runs counter to traditional thinking about the immune system and the brain. "There's been a sort of dogmatic view that the brain is an immunologically privileged site," Johnson says, noting that most scientists believe that the so-called blood-brain barrier keeps most components of the immune system out of the brain, where an immune response could cause serious damage.

Indeed, many scientists offer a far less interesting explanation for the presence of complement proteins in the brains of Alzheimer's patients. They believe that when these patients died, complement proteins floating in the bloodstream managed to slip past a damaged or leaky blood-brain barrier. If correct, the finding of complement proteins would represent merely an artifact and would provide scientists with no new information about Alzheimer's disease, notes Caleb E. Finch, director of the neurogerontology program at USC.

Such views deterred many scientists from exploring the immune system-Alzheimer's link any further. However, the Los Angeles team had completed other research suggesting that brain cells actually manufacture a substance that inhibits or slows the complement system. That finding hinted that the brain did indeed have complement activity going on, and the team began to wonder if brain cells were actually churning out deadly complement proteins. Johnson and Finch began to work on experiments designed to find out more about complement proteins.

In the first study, Finch, Johnson, Giulio M. Pasinetti, and their colleagues created an animal model for Alzheimer's disease. With toxic chemicals or a tiny knife, they damaged nerve pathways in rats' brains to mimic the nerve cell injury seen in the brains of humans who develop Alzheimer's disease. In response to the injury, the rats' brain cells stepped up their production of two types of messenger RNA, molecules that carry the genetic blueprint that tells cells to synthesize specific proteins. In this case, brain cells called microglia cranked out messenger RNA coding for two types of complement proteins.

The rat experiment, reported in the November EXPERIMENTAL NEUROLOGY, suggested that rather than leaking past the blood-brain barrier, complement proteins found in the brain were produced locally by microglial cells. But would the same thing hold true for humans?

To find out, the Los Angeles team obtained brain tissue from seven men and women with Alzheimer's disease who had undergone an autopsy after their death. The group also obtained brain tissue at autopsy from six men and women who were about the same age as the Alzheimer's patients but who did not suffer from dementia.

The researchers discovered that neurons, as well as microglial cells, produced messenger RNA for the two types of complement proteins. The fact that microglia may produce an immune protein didn't come as much of a surprise. These cells are similar to macrophages, immune cells found in the bloodstream, Johnson notes; however, microglia are not considered part of the immune system. Many scientists believe that microglial cells may cruise into the brain during fetal development and do some of the immune system's work once they get there.

However, the scientists did not expect neurons, the information-storing cells of the brain, to produce the messenger RNA for complement proteins. "That's a pretty unexpected finding," Johnson says, noting that this is the first report suggesting that neurons manufacture the immune proteins. Indeed, other scientists express surprise at this result. "I think that finding needs to be confirmed," McGeer says.

The USC team also discovered that brain tissue from the subjects with Alzheimer's disease showed a two- to threefold increase in messenger RNA coding for the two complement proteins. Their results will appear in the November/December NEUROBIOLOGY OF AGING.

If brain cells do manufacture complement proteins, something in the brain may increase the activity, or "turn up the volume," of this system, Finch says. The complement system may act in a constructive fashion in the normal brain, perhaps mopping up the debris left when neurons die of old age, he speculates. But that routine process may go awry in the brain of the Alzheimer's patient, leading to the revved-up and very dangerous complement attack.

The Los Angeles scientists emphasize that the complement system may play only a supporting role in the drama of Alzheimer's disease. Some other factor, perhaps a bad gene, a virus, or an environmental toxin initiates the disease process; complement merely makes the problem worse, Finch suggests. But what gets complement so riled up in the first place?

"That really has been the crux of the matter," says neuroscientist Joseph Rogers at the Institute for Biogerontology Research in Sun City, Arizona. Rogers' team has been studying how and why this system gears up to do its job.

> **"We're not going to say that the complement system has a role yet. But it has the potential."**
> **Steven A. Johnson**

In the rest of the body, the complement system gets involved in the fight against disease in the following way: After a microbe enters the bloodstream, white cells secrete antibodies that attach themselves to the surface of the invader. The first complement molecule, which is called C1Q, recognizes the attached antibody and activates the next complement protein—and so on down the line. The end result is a doughnut-shaped complex that kills the invader. A side effect of this cascade of proteins is the charac-

Tacrine: Reversal of Fortune?

In November 1986, California psychiatrist William K. Summers reported that 12 people with Alzheimer's disease had improved dramatically after taking an experimental drug now known as tacrine (SN: 11/15/86, p.308). These results, including the report that one Alzheimer's patient had resumed playing golf and another had gone back to work, spurred a number of clinical trials designed to test the drug's efficacy.

Since that time, researchers have reported mixed results on tacrine's efficacy and raised concern about liver damage (SN: 3/23/91, p.180).

Now, two new studies add to the growing file on tacrine. They both indicate that this experimental drug does provide benefits to some people with Alzheimer's disease, but the improvements appear modest.

A study in the Nov. 11 JOURNAL OF THE AMERICAN MEDICAL ASSOCIATION (JAMA) suggests that some people with Alzheimer's disease who took tacrine showed significant improvements on tests of memory and intellectual ability. Although there were no dramatic turnarounds, some study participants were able to recognize family members after taking the drug, notes lead researcher Martin Farlow at the Indiana University Medical Center in Indianapolis.

Farlow and his colleagues at 21 U.S. and two Canadian medical centers began their study by recruiting 468 people with mild to moderate Alzheimer's symptoms. The researchers gave patients one of several different doses of tacrine capsules or a placebo. Neither the investigators nor the patients knew which patients received the drug and which got the inactive pills.

After 12 weeks of treatment, the people receiving tacrine showed a statistically significant improvement in the Alzheimer's Disease Assessment Scale, a test designed to measure memory and cognitive abilities. People who received the highest dose of tacrine (80 milligrams per day) showed the greatest improvement on this test, a finding which indicates that response to treatment increases with higher doses, Farlow says.

In addition, tacrine-treated patients exhibited improvement in the Clinical Global Impression of Change, a subjective scale completed by both doctors and family members. This scale helps the researchers gauge any overall change in a patient's ability to function, Farlow says.

About 25 percent of patients taking tacrine developed asymptomatic elevations in liver enzymes circulating in the bloodstream. These enzymes can be an early warning of liver damage, but the problem disappeared when patients stopped taking tacrine, Farlow says. Side effects associated with the drug included nausea, vomiting, diarrhea, and rash.

The second study—presented in part at a U.S. Food and Drug Administration advisory panel meeting last year (SN: 3/23/91, p.180)—shows marginal improvements as a result of tacrine treatment. Kenneth L. Davis of the Mount Sinai Medical Center in New York City and his colleagues studied 215 people with probable Alzheimer's disease who had mild to moderate impairments in memory and cognitive abilities. About half the group received varying doses of tacrine, while the remainder got a placebo. After six weeks, the team discovered that patients getting a tacrine treatment held their own in tests of cognitive ability, while those on placebo got progressively worse. Yet the investigators could detect no across-the-board improvement in people receiving tacrine, says researcher Lon S. Schneider of the University of Southern California in Los Angeles.

Taken together, the two studies add evidence to the belief that tacrine does provide relief, albeit modest, to victims of Alzheimer's disease, comments Gary W. Small at the University of California, Los Angeles. Small believes that Alzheimer's disease has more than one cause. Therefore, tacrine may benefit just a subset of patients, he says in an editorial that appears in the Nov. 11 JAMA.

Alzheimer's disease results in the death of brain cells known to make acetylcholine, a crucial neurotransmitter. Tacrine works by inhibiting an enzyme that breaks down acetylcholine, thus increasing the amount of this neurotransmitter that lingers in brain tissue, Small says. Officials at the Warner-Lambert Co., which manufactures tacrine, remain confident of the drug's ability to clear the remaining regulatory hurdles at the FDA. The agency has yet to approve tacrine, but it has allowed people with Alzheimer's disease to obtain the drug through a limited program, Small says.

— K.A. Fackelmann

---

teristic redness, swelling, and warmth associated with an inflammation.

That classical notion of complement activation runs into trouble when it comes to the brain, where immune components are generally excluded by the blood-brain barrier. However, Rogers' group knew that a few rare substances will trigger the complement system directly, without the help of an antibody.

New research by Rogers' team suggests that the beta amyloid protein itself—the fibrous stuff that forms the core of plaques—can bind with C1Q and set off complement's toxic cascade. They detail their results in the Nov. 1 PROCEEDINGS OF THE NATIONAL ACADEMY OF SCIENCES.

Once activated, the complement system might set up a vicious cycle in which the inflammation produced in the brain tissue leads to more plaque production and cell death, Johnson

says. The whole process may get started many years before the afflicted person notices any symptoms, he adds.

"It could be a very subtle and slow, but continuously destructive mechanism that will be very difficult to track down," Johnson says. "We're not going to say that the complement system has a role yet. But it has the potential."

The USC researchers have yet to prove that human brain cells actually manufacture complement proteins, Johnson cautions. The messenger RNA indicates that the cell has the capacity to make the protein, but it does not show that the cell's protein-making machinery actually spits out these toxic molecules, Finch adds.

"Our work suggests that there is a potential for a destructive mechanism in the brain that is related to this complement system," Johnson says. "That's probably as far as it can be taken [for now]."

If future research does confirm the link between the complement and Alzheimer's disease, it may suggest a change in treatment for the disorder. To date, researchers have not found anything that decisively halts or reverses the symptoms of dementia (see box on page 9).

"I'd hate to speculate a lot," Johnson says. "But if you can track down a problem like this, then I think you're much closer to being able to design a treatment."

Other scientists are already pursuing the treatment angle. Rogers and McGeer believe that therapy with drugs that dampen the inflammatory response might slow the pace of Alzheimer's disease. Their hope is that such medications could prevent the debilitating symptoms altogether if given early in the disease process.

In a letter published in the April 1990 LANCET, McGeer and Rogers put forward the controversial hypothesis that people with rheumatoid arthritis appear to be protected from Alzheimer's disease, perhaps because the anti-inflammatory drugs they take inhibit the complement system.

The investigators looked at several different kinds of prevalence data and found an unexpectedly low prevalence of Alzheimer's disease among people with rheumatoid arthritis, which occurs when the immune system begins a misguided attack on the joints, Rogers notes.

"If complement-mediated attack is one of the mechanisms that would cause damage to the Alzheimer's brain, then you might have some possibility of controlling that by using anti-inflammatory drugs," he adds.

The researchers don't have proof that such a tactic will protect people from the ravages of Alzheimer's disease. The only way to test such a theory is to give anti-inflammatory drugs to people with the disease.

"We have done that," Rogers says, noting that his team has submitted a paper to a journal for publication. "I cannot comment beyond that, except to say that the results appear to be encouraging," he says.

In the end, the hypothesis that the immune system plays a role in the development of Alzheimer's disease remains just that: a hypothesis. And there are other, equally compelling theories that aim to explain this disease.

Johnson and the others know that. Yet they still believe the complement system somehow leads to brain cell death and dementia. They also know that the field remains littered with good ideas that didn't pan out.

"It could turn out to be a big bust in the next several years," Johnson admits. "But I think it's a good lead."

THOUGHT EVOKERS

- What is the major role of the human complement system in relation to the human immune system?

- Describe the first series of symptoms that indicates the onset of Alzheimer's disease.

- Why has it been suggested that the complement system may play only a supporting role in the development of Alzheimer's disease?

4

The Killers All Around

*New viruses and drug-resistant bacteria are reversing
human victories over infectious disease*

By Michael D. Lemonick with J. Madeleine Nash, Alice Park,
Mia Schmiedeskamp, and Andrew Purvis

They can strike anywhere, anytime. On a cruise ship, in the corner restaurant, in the grass just outside the back door. And anyone can be a carrier: the stranger coughing in the next seat on the bus, the college classmate from a far-off place, even the sweetheart who seems perfect in every way. For wherever we go and whatever we do, we are accosted by invaders from an unseen world. Protozoans, bacteria, viruses—a whole menagerie of microscopic pests constantly assaults every part of our body, looking for a way inside. Many others are harmless or easy to fight off. Others—as we are now so often reminded—are merciless killers.

Humanity once had the hubris to think it could control or even conquer all these microbes. But anyone who reads today's headlines knows how vain that hope turned out to be. New scourges are emerging—AIDS is not the only one—and older diseases like tuberculosis are rapidly evolving into forms that are resistant to antibiotics, the main weapon in the doctor's arsenal. The danger is greatest, of course, in the underdeveloped world, where epidemics of cholera, dysentery, and malaria are spawned by war, poverty, overcrowding, and poor sanitation. But the microbial world knows no boundaries. For all the vaunted power of modern medicine, deadly infections are a growing threat to everyone, everywhere. Hardly a week goes by without reports of outbreaks in the U.S. and other developed nations. Some of the latest examples:

• A Royal Caribbean cruise ship on a trip to Baja California returned early

to Los Angeles last week after more than 400 passengers came down with an unidentified intestinal ailment. It may have been the reason one elderly man died. And just a few weeks ago, 1,200 disgruntled passengers were evacuated from the ocean liner *Horizon* in Bermuda because of the threat of Legionnaires' disease. Among customers on previous *Horizon* voyages this summer, there have been 11 confirmed cases of the potentially fatal pneumonia-like illness and 24 suspected cases. At least one victim died.

• A Yale School of Medicine researcher is recovering from a rare and potentially lethal disease called Sabiá virus. Before 1990, the illness was unknown to medicine. Then a woman in the town of Sabiá, Brazil, died from a mysterious virus that had evidently been circulating in local rodents for years before making an assault on humans. Brazilian doctors sent samples to Yale, and a month ago the scientist became infected when he accidentally broke a container holding the virus. Health officials point out that it is not easily passed between humans, but some 80 people who came into contact with the man have been under observation.

• More than 850 people have come down with cholera in southern Russia, and officials fear the disease could erupt into an epidemic. Cholera outbreaks were rare in that part of the world before the breakup of the Soviet Union, but collapsing health services and worsening sanitary conditions have fostered the disease. Shortages of vaccines, meanwhile, have led to an

upsurge in diphtheria in Russia, and health experts have encountered cases of typhoid, hepatitis, anthrax, and salmonella in neighboring Ukraine.

• The notorious flare-up in Gloucestershire, England, of what the press dubbed flesh-eating bacteria alerted people to the dangers of streptococcus-A infections. The common bacteria that cause strep throat generally produce no lasting harm if properly treated, but certain virulent strains can turn lethal. Strep-A infections claim thousands of lives each year in the U. S. and Europe alone.

• Newspaper accounts publicized a startling flare-up of tuberculosis that was first detected last year at a high school in Westminster, California, a middle-class suburb of Los Angeles. The disease was apparently brought in by a 16-year-old Vietnamese immigrant who contracted it in her native country. Nearly 400 young people, or 30% of the school's students, have tested positive for the infection, and at least 12 have a variety of the TB bacterium that is resistant to standard antibiotic treatment. One student has lost part of her lung.

• The *New England Journal of Medicine* reported that the children of Cincinnati suffered an epidemic of pertussis (whooping cough) last year. There were 352 cases (none fatal), compared with 542 cases in the 13 years from 1979 to 1992. The alarming part was that most of the children had been properly vaccinated, suggesting that an unusually hardy strain of the pertussis bacterium might be emerging. Another disturbing statistic: there were more than 6,500 cases nationwide, the largest number in more than 26 years.

• In many parts of the U.S., especially the Northeast, people are already leery of strolling in wooded areas for fear of encountering ticks carrying Lyme disease, a potentially chronic, arthritis-like condition. Now the *Journal of the American Medical Association* has reported on another tick-borne disease, which struck 25 people in Wisconsin and Minnesota, killing two. It is caused by a new variety of the *Ehrlichia* bacterium, which was first detected in humans in 1954. Doctors are concerned because life-threatening *Ehrlichia* infections may be misdiagnosed as Lyme disease or even a bad cold.

A generation ago, no one had ever heard of Lyme or Legionnaires' disease, much less aids. Back in the 1970s, medical researchers were even boasting that humanity's victory against infectious disease was just a matter of time. The polio virus had been tamed by the Salk and Sabin vaccines; the smallpox virus was virtually gone; the parasite that causes malaria was in retreat;

When bacteria began to outwit antibiotics, doctors found themselves retreating in the battle against the germs.

once deadly illnesses, including diphtheria, pertussis, and tetanus, seemed like quaint reminders of a bygone era, like Model T Fords or silent movies.

The first widespread use of antibiotics in the years following World War II had transformed the most terrifying diseases known to humanity—tuberculosis, syphilis, pneumonia, bacterial meningitis, and even bubonic plague —into mere inconveniences that if caught in time could be cured with pills or shots. Like many who went through medical school in the 1960s, Dr. Bernard Fields, a Harvard microbiologist, remembers being told, "Don't bother going into infectious

diseases." It was a declining specialty, his mentors advised—better to concentrate on real problems like cancer and heart disease.

The advent of AIDS demolished that thinking. The sight of tens of thousands of young people wasting away from a virus that no one had known about and no one knew how to fight was a sobering experience—especially when drugs proved powerless to stop the virus and efforts to develop a vaccine proved extraordinarily difficult. Faced with AIDS, and with an ever increasing number of antibiotic-resistant bacteria, doctors were forced to admit that the medical profession was actually retreating in the battle against germs.

The question ceased to be, When will infectious diseases be wiped out? and became, Where will the next deadly new plague appear? Scientists are keeping a nervous watch on such lethal agents as the Marburg and Ebola viruses in Africa and the Junin, Machupo, and Sabiá viruses in South America. And there are uncountable threats that haven't even been named: a virus known only as "X" emerged from the rain forest in southern Sudan last year, killed thousands, and disappeared. No one knows when it might arise again.

A U.S. Army lab in Frederick, Maryland, faced a terrifying situation in 1989 when imported monkeys started dying from a strain of the Ebola virus. After destroying 500 monkeys and quarantining the lab and everyone in it, officials found that this particular strain was harmless to humans. But the episode was dramatic enough to inspire an article in the *New Yorker* magazine—now expanded into a soon-to-be released book called *The Hot Zone*—and work on two competing movies (one of which seems to have collapsed before production).

The Ebola affair and the emergence of AIDS illustrate how modern travel and global commerce can quickly spread disease. Germs once confined to certain regions may now pick up rides to all parts of the world. For example, the cholera plague that is currently sweeping Latin America

arrived in the ballast tanks of a ship that brought tainted water from Asia. And the *New England Journal of Medicine* has reported two cases of malaria in New Jersey that were transmitted by local mosquitoes. The mosquitoes were probably infected when they bit human malaria victims who had immigrated from Latin America or Asia. Writes author Laurie Garrett in a book to be published next month called *The Coming Plague:* "AIDS does not stand alone; it may well be just the first of the modern, large-scale epidemics of infectious disease."

The latest bulletins from the germ front come on top of a long series of horror stories. For years now people have been reading about—and suffering from—all sorts of new and resurgent diseases. As if AIDS were not enough to worry about, there was a rise in other sexually transmitted infections, including herpes, syphilis and gonorrhea. People heard about the victims who died in the Northwest from eating undercooked Jack in the Box hamburgers tainted with a hazardous strain of *E. coli* bacteria. They were told to cook their chicken thoroughly to avoid food poisoning from salmonella bacteria. And last year they saw how the rare hantavirus, once unknown in the U.S., emerged from mice to kill 30 people in as many as 20 states.

All this bad news is undoubtedly having a cumulative impact on the human psyche. The age of antibiotics is giving way to an age of anxiety about disease. It's getting harder to enjoy a meal, make love, or even take a walk in the woods without a bit of fear in the back of the mind. No wonder people pay an unreasonable amount of attention when tabloids trumpet headlines about "flesh-eating bacteria." And no wonder Stephen King's *The Stand*, a TV mini-series based on his novel about a "superflu" that ravages the world's population, earned some of the year's highest ratings.

The odds of contracting a life-threatening infectious disease are still very low—at least in the developed world. But the threats are real and frightening enough to spur medical researchers to redouble efforts to learn more about how the many kinds of microbes cause disease—and how they can be kept at bay.

MICROORGANISMS

It is tempting to think of the tiny pathogens that produce such diseases as malaria, dysentery, TB, cholera, staph, and strep as malevolent little beasts, out to destroy higher forms of life. In fact, all they're trying to do is survive and reproduce, just as we are. Human suffering and death are merely unfortunate by-products.

Plasmodium, a protozoan responsible for malaria, flourishes in the human body, growing inside red blood cells until the cells burst. And without enough red cells to carry oxygen through the body, humans become anemic and can die from renal failure or convulsions. Bacteria, which are considerably smaller than protozoans, generally do their damage indirectly, producing toxins that stimulate the body to mount an immune response. Ideally the immune cells kill the bacteria. But if the bacteria get out of control, their poisons can either kill cells or generate a huge immune reaction that is itself toxic.

In an illness like tuberculosis, the immune system kills the body's own cells in the localized areas where TB germs have taken hold, including the lungs or the bones. With staph or

---

The question ceased to be, When will diseases be gone? and became, Where will the next deadly virus appear?

---

strep, the sheer volume of disease-fighting immune cells can overload blood vessels, ripping tiny tears in the vessel linings; toxins can also damage the vessels directly. Plasma begins to leak out of the bloodstream; blood pressure drops, organs fail, and the body falls into a state of shock. In cholera, bacterial toxins attack intestinal cells, triggering diarrhea, catastrophic dehydration, and death.

Before the coming of penicillin and other antibiotics, bacterial diseases simply ran their courses. Either the immune system fought them off and the patient survived or the battle was lost. But antibiotics changed the contest radically: they selectively killed bacteria without harming the body's cells. For the first time, potentially lethal infections could be stopped before they got a foothold.

Unfortunately, as Columbia University's Dr. Harold Neu observed in the journal *Science*, "bacteria are cleverer than men." Just as they have adapted to nearly every environmental niche on the planet, they have now begun adjusting to a world laced with antibiotics. It didn't take long. Just a year or two after penicillin went into widespread use, the first resistant strain of staph appeared. As other antibiotics came along, microbes found ways to resist them as well, through changes in genetic makeup. In some cases, for example, the bacteria gained the ability to manufacture an enzyme that destroys the antibiotic.

By now nearly every disease organism known to medicine has become resistant to at least one antibiotic, and several are immune to more than one. One of the most alarming things about the cholera epidemic that has killed as many as 50,000 people in Rwandan refugee camps is that it involves a strain of bacterium that can't be treated with standard antibiotics. Relief agencies had to scramble for the right medicines, which gave the disease a head start in its lethal rampage.

Tuberculosis, too, has learned how to outwit the doctors. TB is an unusually tough microbe, so the standard therapy calls for several antibiotics, given together over six months. The length and complexity of the treatment have kept underdeveloped nations from making much progress against even ordinary TB. But now several strains have emerged in the U.S. and other developed countries that can't be treated with common antibiotics.

Even such seemingly prosaic but once deadly infections as staph and strep have become much harder to treat as they've acquired resistance to many standard antibiotics. Both microbes are commonly transmitted from patient to patient in the cleanest of hospitals, and they are usually cured routinely. But one strain of hospital-dwelling staph can now be treated with only a single antibiotic—and public health officials have no doubt that the germ will soon become impervious to that one too. Hospitals could become very dangerous places to go—and even more so if strep also develops universal resistance.

One of medicine's worst nightmares is the development of a drug-resistant strain of severe invasive strep A, the infamous flesh-eating bacteria. What appears to make this variant of strep such a quick and vicious killer is that the bacterium itself is infected with a virus, which spurs the germ to produce especially powerful toxins. (It was severe, invasive strep A that killed Muppeteer Jim Henson in 1990.) If strep A is on the rise, as some believe, it will be dosed with antibiotics and may well become resistant to some or all of the drugs.

Microbes' extraordinary ability to adapt, observes Harvard microbiologist Fields, "is a fact of life. It's written into evolution." Indeed, the end run that many organisms are making around modern antibiotics is a textbook case of Darwin's theory in action (anti-evolutionists, take note). In its simplest form, the theory states that new traits will spontaneously appear in individual members of a given species—in modern terms, mutations will arise in the organisms' genetic material. Usually the traits will be either useless or debilitating, but once in a while they'll confer a survival advantage, allowing the individual to live longer and bear more offspring. Over time, the new survival trait—camouflage stripes on a zebra, antibiotic resistance in a bacterium—will become more and more common in the population until it's universal.

The big difference between animals and bacteria is that a new generation comes along every few years in large beasts—but as often as every 20 minutes in microbes. That speeds up the evolutionary process considerably.

Germs have a second advantage as well: they're a lot more promiscuous than people are. Even though bacteria can reproduce asexually by splitting in two, they often link up with other microbes of the same species or even a different species. In those cases, the bacteria often swap bits of genetic material (their DNA) before reproducing.

They have many other ways of picking up genes as well. The DNA can come from viruses, which have acquired it while infecting other microbes. Some types of pneumococcus, which causes a form of pneumonia, even indulge in a microbial version of necrophilia by soaking up DNA that spills out of dead or dying bacteria. This versatility means bacteria can acquire useful traits without having to wait for mutations in the immediate family.

The process is even faster with antibiotic resistance than it is for other traits because the drugs wipe out the resistant bacterium's competition. Microbes that would ordinarily have to fight their fellows for space and nourishment suddenly find the way clear to multiply. Says Dr. George Curlin of the National Institute of Allergy and Infectious Diseases: "The more you use antibiotics, the more rapidly Mother Nature adapts to them."

Human behavior just makes the situation worse. Patients frequently stop taking antibiotics when their symptoms go away but before an infection is entirely cleared up. That suppresses susceptible microbes but allows partially resistant ones to flourish. People with viral infections sometimes demand antibiotics, even though the drugs are useless against viruses. This, too, weeds out whatever susceptible bacteria are lurking in their bodies and promotes the growth of their hardier brethren. In many countries, antibiotics are available over the counter, which lets patients diagnose and dose themselves, often inappropriately. And high-tech farmers have learned that mixing low doses of antibiotics into cattle feed makes the animals grow

larger. (Reason: energy they would otherwise put into fighting infections goes into gaining weight instead.) Bacteria in the cattle become resistant to the drugs, and when people drink milk or eat meat, this immunity may be transferred to human bacteria.

Because microbial infections keep finding ways to outsmart antibiotics, doctors are convinced that vaccines are a better way to combat bacterial disease. A vaccine is usually made from a harmless fragment of microbe that trains the body's immune system to recognize and fight the real thing. Each person's immune system is chemically different from everyone else's, so it's very difficult for a bacterium to develop a shield that offers universal protection. Diphtheria and tetanus can be prevented by vaccines if they are used properly. A vaccine against the pneumococcus bacterium has recently come out of the lab as well, and scientists expect to test one that targets streptococcus A within a year.

VIRUSES

Unlike bacteria and protozoans, which are full-fledged living cells, capable of taking in nourishment and reproducing on their own, viruses are only half alive at best. They consist of little more than a shell of protein and a bit of genetic material (DNA or its chemical cousin RNA), which contains instructions for making more viruses—but no machinery to do the job. In order to reproduce, a virus has to invade a cell, co-opting the cell's own DNA to create a virus factory. The cell—in an animal, a plant, or even a bacterium—can be physically destroyed by the viruses it is now helplessly producing. Or it may die as the accumulation of viruses interferes with its ability to take in food.

It is by killing individual cells in the body's all-important immune system that the AIDS virus wreaks its terrible havoc. The virus itself isn't deadly, but it leaves the body defenseless against all sorts of diseases that are. Other viruses, like Ebola, kill immune cells too, but very quickly; the dead cells form massive, deadly blood clots. Still others, hantavirus, for example, trigger a powerful reaction in which

immune cells attack both the invading virus and the host's healthy cells.

Unlike bacteria and protozoans, viruses are tough to fight once an infection starts. Most things that will kill a virus will also harm its host cells; thus there are only a few antiviral drugs in existence. Medicine's great weapon against viruses has always been the preventive vaccine. Starting with smallpox in the late 1700s, diseases including rabies, polio, measles, and influenza were all tamed by immunization.

But new viruses keep arising to challenge the vaccine makers. They may have gone undetected for centuries, inhabiting animal populations that have no contact with mankind. If people eventually encounter the animals—by settling a new part of the rain forest, for example—the virus can have the opportunity to infect a different sort of host.

Scientists believe Ebola virus made just that kind of jump, from monkeys into humans; so did other African viruses such as Marburg and the mysterious X that broke out in Sudan. And many more are likely to emerge. "In the Brazilian rain forest," says Dr. Robert Shope, a Yale epidemiologist, "we know of at least 50 different viruses that have the capacity of making people sick. There are probably hundreds more that we haven't found yet."

Viruses like Ebola and X are scary, but they're too deadly to be much of a threat to the world. Their victims don't have much of a chance to infect others before dying. In contrast, HIV, the AIDS virus—which may have come from African primates as early as the 1950s—is a more subtle killing machine and thus more of an evolutionary success. An infected person will typically carry HIV for years before symptoms appear. Thus, even though HIV doesn't move easily from one human to another, it has many chances to try. Since the first cases were reported in the late 1970s, HIV has spread around the world to kill perhaps a million people and infect an estimated 17 million.

It isn't just new viruses that have doctors worried. Perhaps the most ominous prospect of all is a virulent strain of influenza. Even garden-variety flu can be deadly to the very old, the very young, and those with weak immune systems. But every so often, a highly lethal strain emerges—usually from domesticated swine in Asia. Unlike HIV, flu moves through the air and is highly contagious. The last killer strain showed up in 1918 and claimed 20 million lives—more than all the combat deaths in World War I. And that was before global air travel; the next outbreak could be even more devastating.

Vaccines should, in theory, work just as well for new varieties of disease as they do for old ones. In practice, they often don't. An HIV vaccine has proved difficult to develop because the virus is prone to rapid mutations. These don't affect its deadliness but do change its chemistry enough keep the immune system from recognizing the pathogen.

Creating a vaccine for each strain of flu isn't exactly simple either. First, says Yale's Shope, "we have to discover something new is happening. Then we have to find a manufacturer willing to make a vaccine. Then the experts have to meet and decide what goes into the vaccine. Then the factory has to find enough hens' eggs in which to grow the vaccine. There are just a lot of logistical concerns."

People are partly to blame for letting new viruses enter human populations. Says Dr. Peter Jahrling, senior research scientist at the U.S. Army Medical Research Institute of Infectious Diseases: "If you're a monkey imported from the Philippines, your first stop when you hit this country is a quarantine facility. If you're a free-ranging adult human being, you just go through the metal detector and you're on your way."

Sometimes environmental changes help microbes move from animals to humans. Lyme disease, a bacterial infection, was largely confined to deer and wild mice until people began converting farmland into wooded suburbs—which provided equally good habitats for the animals and the bacteria-infested ticks they carry and also brought them into contact with large numbers of people. The mice that transmit the hantavirus often take refuge in farmers' fields, barns, and even homes. Air-conditioning ducts create a perfect breeding ground for Legionnaires' disease bacteria. Irrigation ditches and piles of discarded tires are ideal nesting spots for the *Aedes aegypti* mosquito, carrier of dengue and yellow fevers; imported used tires have already brought the Asian tiger mosquito, also a carrier of dengue, into the U.S.

Clearly there is no way to prevent human exposure to microbes. But the risks can be reduced. To minimize bacterial resistance, for example, doctors can be stingier with antibiotics. "We've been careless," says Dr. Robert Daum, a University of Chicago pediatrician. "Every childhood fever does not require antibiotics." Nor does a healthy farm animal.

Most important is increased vigilance by public-health authorities. The faster a new microbe can be identified and its transmission slowed, the less likely a small outbreak will turn into an epidemic. Unfortunately, the trend has been in the other direction. "Even in the U.S.," says Thomson Prentice of the World Health Organization in Geneva, "disease-monitoring expertise has been lost, either through cost-cutting or reduced diligence. If some of the edge has been lost in the U.S., imagine how poorer countries have reacted."

American health officials are convinced that their information-gathering network must be strengthened. That has begun to happen under a new program that will, among other things, increase the surveillance of new microbes and educate both health workers and the public about how to deal with emerging diseases.

An all-out effort to monitor diseases, vaccinate susceptible groups, improve health conditions around the world, develop new drugs, and get information to the public would be enormously expensive. But the price of doing nothing may be measured in millions of lost lives. Doctors are still hopeful but no longer overconfident. "I do believe that we're intelligent enough to keep ahead of things," says epidemiologist Shope. Nonetheless, neither he nor any of his colleagues will ever again be foolish enough to declare victory in the war against the microbes.

THOUGHT EVOKERS

- Describe the differences between the three major unicellular agents of disease: viruses, bacteria, and protozoa.

- Explain how microbes are moved from host to host.

- Which are the world's deadliest microbe-caused diseases?

Tinkering
with
Destiny

5

By Shannon Brownlee, Gareth G. Cook, and Viva Hardigg

The last thing Joey Paulowsky needs is another bout with cancer. Only 7 years old, the Dallas native has already fought off leukemia, and now his family worries that Joey could be hit again. The Paulowsky family carries a genetic burden, a rare form of inherited cancer of the thyroid. Deborah, his mother, found a lump in her neck six years ago, and since then one family member has died of the cancer and 10 others have had to have their thyroids removed. "Do I have cancer?" Joey asks his mother. "Will it hurt?" The Paulowskys will know the answer next month, when the results of a genetic test will show whether their son carries the family's fateful mutation.

Joey is too young to know that he is participating in a medical revolution, one that will change the practice of medicine as profoundly as the invention of the microscope or the discovery of antibiotics. The snippet of DNA that will determine his fate was identified barely a year ago. Now, this mutant gene, along with the more than 150 others that have been captured thus far, are making it possible for doctors to peer into their patients' medical futures. Today, at least 50 genetic tests for hereditary diseases are available; by the turn of the century, DNA tests are almost certain to be a standard part of medical exams. From a single sample of a patient's blood, doctors will be able to spot genetic mutations that signal the approach not only of rare hereditary diseases, such as the thyroid cancer that stalks the Paulowsky clan, but

also the common killers, including breast cancer, heart disease, and diabetes—and defeat them.

For all its promise, the ability to glimpse the future will not come without costs. Knowing a patient's genetic predispositions will be central to preventive medicine, a keystone of health care reform; a physician will one day be able to advise the young adult at risk for high blood pressure in middle age to cut down on salt long before the appearance of symptoms. But for many other inherited ailments, a genetic test offers a Faustian bargain. For example, women who are mem-

---

"There are two struggles. One is the disease itself. The other is not knowing."

---

bers of families at high risk for breast cancer will soon be able to undergo a genetic test for the breast cancer gene (known as BRCA1), which is responsible for as many as 1 in 10 of the 180,000 cases diagnosed in the United States each year. Those who are found to be free of the mutation will be spared the dread of the unknown. But those who do carry the mutation can only hope that self-examinations and mammograms catch the tumor early enough. Some women opt to have their healthy breasts removed, though there is little evidence that even this will prevent the cancer.

The rate of change is unlike anything medicine has witnessed before, as researchers fish genes out of cells at a dizzying rate. Last year saw the dis-

covery of more than a dozen mutations responsible for diseases ranging from Alzheimer's to hyperactivity to colon cancer. Almost as soon as a gene is discovered, commercial laboratories are ready to offer a genetic test—a pace that threatens to outstrip both physicians' and patients' abilities to make sense of the information. Couples forgo having children after misunderstanding the result of genetic tests, while patients who carry mutations can lose their insurance under the current health care system. Scientists, genetic counselors, and ethicists are racing to develop guidelines for the new age of genetic medicine, while families ponder the dilemmas presented by an incomplete medical revolution. "If we screw this up," says Francis Collins, director of the National Center for Human Genome Research, "I don't think the public will conclude that this was a useful revolution."

THE FAMILY

With cancer in my genes, is it safe to marry and have children?

On the surface, the five daughters of the Kostakis* family seem alike. They all have the same mass of curly hair, the same lively manner of speaking and the same propensity to start an argument in English and finish it in Greek. Yet an invisible and devastating distinction divides them: Two of the sisters inherited their mother's gene for a deadly form of colon cancer: the other three, like their brother, did not.

The family immigrated in 1977 from an impoverished village in Greece,

---

The names Kostakis and Smith have been created to protect patients' privacy.

intent on forging a new and better life in America and oblivious to the genetic cargo that menaced their dream. "If I had known I carried the disease, I never would have gotten married and had children," says their mother, Eleni, in Greek. It was only when she went for a tubal ligation at age 39 that clues to the family's legacy began to emerge.

Eleni told the doctor that her father had died at age 28, with a stomach tumor so large she could feel it with her hand, and that her older brother had undergone mysterious surgery he didn't like to talk about. The gynecologist immediately scheduled a colonoscopy—a jarring, uncomfortable procedure done under sedation. The eye of the colonoscope revealed the warning signs: benign polyps numbering in the hundreds or even thousands—harbingers of familial adenomatous polyposis colon cancer, a disease that can strike in some families as early as the teens.

Soon after Eleni had her colon removed, her younger brother was hit. He refused to see a doctor until a week before he died, at age 43, in a manner hauntingly reminiscent of her father's death. "He died because he chose to," she says, shaking her head and balling her hands into tight fists in her lap. Eleni vowed that her children would know what they were up against. That meant yearly colonoscopies for the Kostakis siblings, each of whom had a 50-50 chance of inheriting the ancestral cancer gene. Two of the sisters had the telltale polyps and each had a portion of her colon removed at age 16.

Now, the family knows definitively who in the new generation has been spared. In 1991, researchers discovered the gene that when mutated is responsible for adenomatous polyposis, and like 90 percent of those offered the gene test at the Johns Hopkins School of Public Health, Eleni's other four children opted to take it as soon as it was offered. "Anything that might stop those colonoscopies was a great relief," says Maria, 26, the eldest. But when it came time to find out the results, everyone was on edge. Petros, 25,

remembers his heart pounding in the doctor's office before the genetic counselor told him and his sisters the good news: all negative. "Finding out was the biggest relief of my entire life," Petros recalls. "It was better than making a million dollars, better than sex—almost." Two weeks ago, a 14-year-old cousin received the same happy tidings.

Maria felt a similar surge of freedom. "There are two struggles in a cancer family," she says. "One is the disease and one is not knowing. You have inside this *anchos*, we call it in Greek. It's like a bugging feeling, this worry." For known carriers of the gene, knowledge has its price, especially when it comes to the agonizing issue of parenthood. Katerina, the youngest of the family, has recovered well from her colectomy and has resumed the life of a typical high school senior. But the future haunts her. She clutches a small pillow against her stomach and fights back tears as she says, "The thing that scares me the most is if I found out my baby had it, I would abort it."

As devastating as their illness is, the Kostakises are among the lucky ones. Not only is there a gene test for their type of cancer but, for those who carry it, the telltale polyps show where and when to operate. By contrast, Huntington's disease, the devastating neurological degeneration that killed

---

"We had the sense we were close to the secret of cancer. Then we thought, 'Now What?' "

---

folk singer Woody Guthrie at 55, has a test but no cure. About 85 percent of people at risk for Huntington's have declined the test, preferring to live in blissless ignorance.

For families suffering from many other familial cancers, the *anchos* remains, and each year is a waiting game. Margaret Todd, 66, of Towson, Md., watched her parents die of colon cancer, and she herself has had malignant polyps removed. Her family's tumors arise from hereditary nonpolyposis colon cancer, which often fails to

produce warning polyps and cannot yet be detected through genetics, even though the genes responsible were discovered last year. If there were a test, Todd's four children would know whether they had to endure the discomfort and expense of the annual colonoscopies, which cost about $1,000 apiece. She says, "We keep praying for a gene test and they keep saying it might come this year."

THE SCIENTISTS
Where are the seeds of cancer and other genetic diseases?

Ray White knows which members of the Smith family will get colon cancer, but most of the Smiths still do not. It is a position relished by neither the Smiths nor White, who heads the Huntsman Cancer Institute at the University of Utah. White's team of researchers discovered the gene responsible for the Smiths' colon cancer three years ago. Since then, he says, "they have been beating on our door for the information." But for a variety of reasons, all aimed at protecting the Smiths, the researchers were not ready until this June to share the news with the family.

White did not set out to be the keeper of such grave information when he arrived in Salt Lake City in 1980, armed only with a new genetic technique of his own invention and the desire to hunt disease-causing genes. This was the dawn of the genetic age, when White and other scientists would finally begin to make sense of more than 4,000 heritable diseases. In the previous two decades, scientists had unraveled the mystery of inheritance, showing that it is governed by DNA, the genetic material contained in the nucleus of each of the body's cells. In 1980, White and a colleague at the Massachusetts Institute of Technology devised a means for searching through DNA, infinitesimal section by infinitesimal section. By comparing the length of the same section from one family member to another, researchers now had a way to zero in on the approximate location of genes corresponding to inherited diseases.

In 1982, White aimed his sights at the mangled DNA responsible for adenomatous polyposis colon cancer, or APC. APC accounts for only 1 percent of all colon cancer cases, but it snakes through the family trees of half a dozen Mormon clans in Utah. Mormons proved perfect subjects for White's team: They keep meticulous genealogies, and many families bear as many as 10 to 12 children, giving an inherited disease a chance to show up in each generation. The researchers spent a decade flying around the state collecting blood samples, often at family reunions. "On holidays, I can figure I'm going to be drawing blood," says Utah geneticist Ken Ward. In 1991, White's team, along with a group led by Bert Vogelstein at Johns Hopkins Oncology Center in Baltimore, announced simultaneously that they had nailed the colon cancer gene.

Yet even in their euphoria the researchers knew they could not simply call the Smiths, a clan of some 2,000 to 3,000, and blurt out who was safe and who wasn't. First, they had to find the precise mutations in the gene that ran through their broad family tree—then estimate the odds that an individual carrying a mutation would actually be hit with cancer. The researchers also worried that family members could lose their insurance once it was discovered they harbored the defective gene. One man's insurance was canceled simply because he participated in the genetic study—though it turned out he did not have the gene. The researchers had to be certain of just what their scientific findings meant in practical terms to families at risk for cancer. This June, the first letters arrived in Smith mailboxes, offering a chance to learn their genetic legacy.

Researchers around the world will find themselves in a similar fix with each new gene they uncover. A gene found in 1991 by Utah geneticist Jean Marc LaLouel now appears to bestow some of its owners with high blood pressure while leaving others vulnerable to certain complications of pregnancy. Earlier this year, researchers at Thomas Jefferson University in

Philadelphia announced they are close to the long-sought gene responsible for manic-depression. This summer, a French lab fetched up a gene that leads to melanoma, a deadly form of skin cancer, while geneticists at the National Institutes of Health announced that p53, a gene notorious for its role in more than half of all cancers, may also make mischief in the cells lining the arteries, thus contributing to heart disease. But while these bits of DNA are already opening new avenues to combatting and perhaps even preventing illness, cures are probably years, if not decades, away.

Until then, researchers want to put their newfound genes to use, spotting diseases as early as possible. Johns Hopkins University's David Sidransky is confident that the ubiquitous p53 gene can serve as a red flag for tumors of the mouth and bladder. Sidransky analyzed a urine specimen saved from Hubert Humphrey in 1967 and found mutant copies of p53. If the gene had been known at the time, it might have alerted doctors to the bladder cancer that would kill the senator a decade later. Another cancer gene, called RAS, can warn of impending lung cancer; RAS and the APC gene in stool samples may one day alert doctors to as many as 80 percent of colon cancers—even among the 150,000 cases

---

"We have this fantasy that if we just have enough information we can control events."

---

that arise each year in people with no familial history. For cancer, more than almost any other illness, such early warning systems are desperately needed. "We are going to figure out how to identify precancerous lesions and get rid of them," says White, "before they turn into full-blown cancer."

THE COUNSELOR

Are people capable of understanding genetic risk?

The day a cure for cancer is discovered will be the day Barbara Biesecker's job is made easier.

Biesecker is a genetic counselor, one of the medical messengers who are trained in both psychology and human genetics—and who bear the responsibility for making sure that people like the Smiths and the Kostakises understand the implications of their genetic tests. The result of these tests is rarely a simple thumbs up or down. More often, the best a genetic test can offer is a degree of risk, and conveying the meaning of risk is no easy task.

Weighing the odds. Even when people comprehend the numbers (and many cannot), they find uncertainty psychologically troubling. When confronted with a 50 percent risk, many patients conclude their chances are either zero or 100 percent, says Biesecker, head of genetic counseling at the National Center for Human Genome Research in Bethesda, Md. "It's either, 'I've got it' or 'I don't.'"

The flood of new gene discoveries has left genetic counselors unsure of how to proceed in the new age of medical genetics. The counselors' credo requires that they help clients come to their own decisions, but they are concerned about the increasing numbers of people—particularly expectant couples—who demand the newest genetic tests even when there is little evidence of medical risk. "We have this fantasy in this country that if we have enough information, we can control events," says Biesecker.

This desire for information is especially troubling when parents want their children tested for diseases whose symptoms will not appear for many years. "Parents think they want their children tested, but what they really want to hear is that their child does not have the disease," says Randall Burt, a gastroenterologist at the Veterans Hospital in Salt Lake City. They haven't thought through what a positive result will mean to them.

Few of the new genes have raised more nettlesome issues than BRCA1, the breast cancer gene. Geneticists estimate that 1 in 200 women may carry mutations in this gene; millions will no doubt want to be tested. There are only about 1,200 genetic counselors in the country, not nearly enough to handle the job of deciphering the results. For example, a positive test means a woman's daughters could also harbor the mangled gene. A negative test for BRCA1, on the other hand, does not mean a woman has entirely dodged the bullet; she still faces the possibility of getting other forms of breast cancer. And for women who have watched their mothers and sisters die of breast cancer, a negative test sometimes leads to "survivor guilt," feelings similar to those of victims who escaped the Holocaust.

With a shortage of genetic counselors, the task of sorting through such dilemmas will fall to doctors or even to commercial testing services. "What will happen when this is in the hands of primary care physicians scares me a lot," says Biesecker. "Are physicians prepared to draw the line between giving advice about treatment and advising a couple whether or not to have children?"

THE DOCTOR

Are doctors ready to practice medicine in the genetic age?

How physicians will use the fruits of the genetic revolution is a major concern of Gail Tomlinson, a pediatric oncologist at the University of Texas Southwestern Medical Center in Dallas. Tomlinson knows firsthand both the power and the peril of genetic tests: As 7-year-old Joey Paulowsky's physician, she will use the test results due next month to decide whether the child's thyroid should be removed.

But few of Tomlinson's colleagues are as well trained as she is in the complexities of medical genetics. Unlike genetic counselors, who learn the tricky business of informing without recommending, most doctors are taught to tell patients what is best for them. That is not easy when all a test offers is a measure of risk. For example, among Tomlinson's patients are families afflicted with Li Fraumeni syndrome, a rare hereditary mutation in the p53 gene that can bring on a bewildering variety of cancers. But until there are better ways to detect the tumors early enough to stop them, the test for p53, says Tomlinson, "may not do patients a lot of good. You can't fix the defective gene, and you can't catch many of the cancers."

Eventually, genetic tests will belong in the hands of the doctors on the front lines of medicine, says Tomlinson. "But things are moving so fast. We know how to clone genes, but we don't know how to talk about it to patients. Just how to apply this predictive testing hasn't been worked out."

Until it is, some physicians will view genetic tests with a mixture of unease and distrust. Sandra Byes, an oncologist at the University of Utah Medical Center, is uncertain whether she will advise her patients to take a genetic test for the breast cancer gene once it is discovered: "It's a lot more comforting to say to a patient, 'You are from a high-risk family' than 'You have the gene, try not to let it ruin your life.'"

Gambling on prevention

Linda Abbate and Nadine Antin-Colla share a bond that links hundreds of thousands of American women: Both have watched several close relatives struggle with breast cancer. Abbate's two older sisters have undergone mastectomies. Antin-Colla's mother and aunt died of the disease. Their family histories make Abbate and Antin-Colla candidates for a controversial new study to see whether the drug tamoxifen can protect healthy women against breast cancer.

The tamoxifen project is one of several trials aimed at stopping cancer and other inherited diseases before they ever develop. But the practice has also raised ethical concerns about treating healthy people who may or may not get a disease with drugs that have potentially dangerous side effects.

Reason to believe

Studies of women already diagnosed with breast cancer find that tamoxifen greatly reduces the risk of dying from the spread of the disease and lowers the chances of a tumor's cropping up in the unaffected breast. Yet critics, including the National Women's Health Network, argue that making the leap from treating a disease to preventing its occurrence in healthy patients is risky. Since the genetic culprit that causes breast cancer has not yet been identified, there is no simple test to determine which women actually have inherited an increased likelihood for the disease; risk is assessed largely on the basis of family history. And tamoxifen is not without

dangers of its own: Studies show that women taking the drug have 7 1/2 times the average risk of dying from uterine cancer.

Such trials with preventive drugs are controversial for other reasons as well. A new drug now being tested for its abilities to prevent the onset of prostate cancer, for example, may have the perverse effect of masking signs of the illness. In an ongoing prostate cancer prevention trial, men are being randomly assigned to take either a placebo or the drug finasteride, which is effective in treating an enlarged prostate. But studies show that finasteride can reduce the levels of a protein in the blood called prostate specific antigen, which is frequently elevated in men with prostate cancer and so is used as a marker in diagnosing the disease. Thus a drug with unproved effectiveness may actually hinder early diagnosis of the disease.

The uncertainty surrounding the new drugs makes choosing treatments as much a matter of psychology as of medicine. For Abbate, the possibility of avoiding breast cancer—second only to lung cancer as the most lethal malignancy for American women—is worth the chance of getting a less common and more easily cured disease, such as cancer of the uterus. Antin-Colla, on the other hand, dropped out of the study after learning of the link between tamoxifen and uterine cancer. "My own personal history has been a preoccupation with worry about breast cancer," she says. "I just want to worry about only what I'm used to worrying about."

Rita Rubin

THE ENTREPRENEUR

Is American society on the verge of a genetic gold rush?

In 1989, business was so bad at the DNA lab at Integrated Genetics that the company was close to shutting it down. Now, the respected Framingham, Mass., lab boasts nearly one fifth of the $8 million market in DNA tests for inherited diseases, a fledgling industry poised to take off. Industry analysts foresee a $500 million market in genetic tests in the next decade.

The potential for profits is bringing entrepreneurs like IG President Elliott Hillback face to face with a marketplace filled with moral pitfalls. With the discovery of each disease-causing gene, a new and lucrative market opens, but it can take years before anyone understands the medical seriousness of any given genetic mutation (and some genes can have hundreds of mutations). Last year, for example, more than 30,000 tests were performed to detect one or more of some 350 mutations in the gene behind cystic fibrosis, a heritable and often fatal lung disease. Yet scientists are only now discovering that some mutations actually cause none of the classic symptoms. The danger is that some labs will make a test available before the results can be meaningfully interpreted.

Private codes. Critics also fear for the patient's right *not* to know the contents of his genes. Today, the final responsibility for explaining the untoward consequences of a test—such as losing health insurance—lies with the laboratory itself. But turning away patients means turning away profits.

The two government agencies charged with overseeing the burgeoning genetics market, the Food and Drug Administration and the Health Care Financing Administration, have been slow to step in. Most of the companies that manufacture the chemical tools of the genetics trade have not bothered to submit their products to the FDA for approval as required by law. Labs serve a vital role in making tests cheaper and more accurate. But HCFA, charged with ensuring test accuracy, "has dragged its feet," according to Neil Holtzman, a member of a Human Genome Project working group investigating genetic testing. This is in part because the agency lacks inspectors with the necessary genetic training. For now, there is nothing—beyond the lab director's personal ethics—to keep a lab from introducing new tests, regardless of their potential for misuse.

A few companies are already pushing the boundaries. Genica Pharmaceuticals of Worcester, Mass., for example is offering a test for a gene linked to Alzheimer's disease even though the Alzheimer's Association has said that too little is known about the gene and its mutations to interpret test results properly. Genica President Robert Flaherty admits that "it is not a definitive test," but believes that under the right circumstances "it can be very useful."

While Hillback worries that a few reckless labs might cause a public backlash against the entire enterprise, it is ultimately not the responsibility of commercial establishments to set boundaries. Americans must decide where to draw the ethical line. And wherever they finally choose to draw it, Hillback and his colleagues know that there will be no turning back. "It's much easier to say that society as a whole is not ready for a test," says James Amberson, a vice president at Dianon Systems, a testing company. "But it's awfully hard to look a mother in the eye and say no."

THOUGHT EVOKERS

- Discuss the legal and moral consequences that might arise from "genetic tinkering."

- Explain the range of emotions involved with the discovery of BRCA1—the breast cancer gene.

- Explain what is meant by the phrase "the patient's right not to know" regarding genetic counseling.

DNA Tests Available Now

Disease	Description	Incidence	Cost
Adult polycystic kidney diseases	Multiple kidney growths	1 in 1,000	$350
Alpha-1-Antitrypsin deficiency	Can cause hepatitis, cirrhosis of the liver, emphysema	1 in 2,000 to 1 in 4,000	$200
Charcot-Marie-Tooth disease	Progressive degeneration of muscles	1 in 2,500	$250–$350
Familial adenomatous polyposis	Colon polyps by age 35, often leading to cancer	1 in 5,000	$1,000
Cystic fibrosis	Lungs clog with mucus; usually fatal by age 40	1 in 2,500 Caucasians	$125-$150
Duchenne/Becker muscular dystrophy	Progressive degeneration of muscles	1 in 3,000 males	$300–$900
Hemophilia	Blood fails to clot properly	1 in 10,000	$250-$350
Fragile X syndrome	Most common cause of inherited mental retardation	1 in 1,250 males; 1 in 2,500 females	$250
Gaucher's disease	Mild to deadly enzyme deficiency	1 in 400 Ashkenazi Jews	$100–$150
Huntington's disease	Lethal neurological deterioration	1 in 10,000 Caucasians	$250-$300
"Lou Gehrig's disease" (ALS)	Fatal degeneration of the nervous system	1 in 50,000, 10% familial	$150–$450
Myotonic dystrophy	Progressive degeneration of muscles	1 in 8,000	$250
Multiple endocrine neoplasia	Endocrine gland tumors	1 in 50,000	$900
Neurofibromatosis	*Café au lait* spots to large tumors	1 in 3,000	$900
Retinoblastoma	Blindness; potentially fatal eye tumors	1 in 20,000	$1,500
Spinal muscular atrophy	Progressive degeneration of muscles	7 in 100,000	$100–$900
Tay-Sachs disease	Lethal childhood neurological disorder	1 in 3,600 Ashkenazi Jews	$150
Thalassemia	Mild to fatal anemia	1 in 100,000	$300

Tests of the Future

Disease	Description	Incidence	Cost
Alzheimer's	Most likely multiple genes involved	4 million cases	Not available
Breast cancer	Five to 10% of all cases are thought to be hereditary	2.6 million cases	Not available
Diabetes	Most likely multiple genes involved	13–14 million cases	Not available
Nonpolyposis colon cancer	Several genes cause up to 20% of all cases	150,000 cases per year	Not available
Manic-depression	Most likely multiple genes involved	2 million cases	Not available

Source: Helix: National Directory of DNA Diagnostic Laboratories, Children's Hospital and Medical Center,

Seattle, Wash. For more information: Alliance of Genetic Support Groups, (800) 336-4363.

6

Antibiotic Resistance

Mechanisms preventing antibiotics from killing bacteria are appearing much faster than ways to control resistance

By Carlos F. Amábile-Cuevas, Maura Cárdenas-García, and Mauricio Ludgar

There is a new monster in American books and films. This monster is not a giant lizard or ape, nor is it a body-snatching alien or an undead creature back to claim its legacy. The new monster is a microscopic agent of disease that is stirred out of its exotic—and remote—locale by people who are unwilling to leave well enough alone. The exotic pathogen is transported back to an urban American center where its lethal and unstoppable effects travel swiftly through the population. Thousands of people die before the lone, outcast scientist finds the pathogen's weak spot, produces a cure, and banishes the pathogen, at least for the time being, to the safe confines of the annals of medical history.

The plot has its roots in the recent past, where unusual viruses, such as HIV and hantavirus, and deadly bacteria, such as the so-called flesh-eating bacterium, seem to have appeared suddenly and forcefully on the American public health scene. But in fact many of the newest microbial enemies are neither strange nor exotic.

The authors work at LUSARA, a research institution devoted to the study of antibiotic resistance and other related molecular biological problems. Carlos Amábile-Cuevas obtained his D.Sc. from the Center for Research and Advanced Studies in Mexico City. He is founder and head of LUSARA, now 10 years old. Maura Cárdenas-García received her B.Sc. and M.Sc. degrees from the National Autonomous University of Mexico and joined the research team last year. Mauricio Ludgar is a high school and college science teacher who has been involved with LUSARA for several years. Address for Amábile-Cuevas: LUSARA, Apartado Postal 102-006, Mexico City 08930, Mexico.

Some of the most dangerous recent public health threats are shockingly familiar—diseases such as tuberculosis and typhoid fever, evocative of a former era when infectious bacteria were often deadly killers.

The reason these and other infectious agents are making a comeback is that they are no longer killed by the drugs that for the past 50 years have kept them at bay; they have become resistant to antibiotics. Antibiotic resistance has also made potential killers out of bacteria that formerly were not much of a threat. And some bacteria are resistant to not just one, but to several different antibiotics, making it difficult for clinicians to hit on a drug regime that will fend them off.

This state of affairs results from an involuntary experiment in evolution and natural selection. The indiscriminate and, in retrospect, reckless use of antibiotics has selected for increasingly resistant bacteria. Each instance of antibiotic use kills off only susceptible bacteria. Some small population of resistant organisms is left behind, free to multiply and to pass on the resistance genes to other individuals until eventually resistant organisms outnumber antibiotic-sensitive ones. Within just a few decades, medicine has created these new antibiotic-resistant monsters from what used to seem like a manageable nuisance. What's more, mechanisms of resistance can be passed around from one bacterial cell to another, and even between cells of different (or distantly related) bacterial species. Bacteria also achieve some measure of resistance by forming colonies, which must be taken into account by any researchers hoping to revive old antibiotics and discover new ones.

Home-Grown Monsters

Historically, outbreaks of infectious disease have been associated with the poor sanitation conditions of crowded urban areas. So it is ironic that one of the best incubators for today's growing number of resistant bacteria is not a stagnant pond or an open sewer, or any of the other traditional breeding grounds for microbes. Many of today's resistant bacteria can be found and spread in the hospital. Often, bacterial strains resistant to one or two drugs enter the hospital via the patients and are spread within the hospital on the hands of hospital personnel, through the air, and on hospital surfaces. Clinical isolates of *Pseudomonas aeruginosa*, resistant

> ... one of the best incubators for today's growing number of resistant bacteria is not a stagnant pond or an open sewer, or any of the other traditional breeding grounds for microbes. Many of today's resistant bacteria can be found and spread in the hospital.

because their cell walls no longer admit antibiotics, are a frequent cause of nosocomial infections, those acquired within the hospital. *Pseudomonas* commonly causes respiratory and urinary tract infections.

Because of the concentrated exposure to antibiotics, hospital strains of bacteria "collect" resistance determi-

Reprinted by permission of *American Scientist*, Journal of Sigma Xi, The Scientific Research Society.

nants to drugs and disinfectants at a higher rate than strains found outside of the hospital. As a result, hospitals not only contain more resistant bacteria than other milieus, but they contain more multiply resistant strains than other sites. Bacteria such as *Klebsiella, Serratia, Proteus,* and *Enterobacter,* which formerly yielded to antibiotic treatments, have become an increasing cause of hospital-acquired infection, especially dangerous because they are multiply resistant to antibiotics. Hospitals, however, are not the exclusive sites of resistance outbreaks, which have been observed in the general population for the past 20 or so years.

Among the most important causes of community-acquired infections is the almost-garden-variety bacterium *Escherichia coli.* This bacterium is found copiously in the human gut, where it aids digestion. Recently, strains of *E. coli* that are resistant to several different drugs have been found. Multiresistant *E. coli* is a common cause of failure of antibiotic therapy in outpatient care and a reservoir of antibiotic-resistance genes.

The first well-documented bacterial outbreak involving multiply resistant bacteria was an epidemic of typhoid fever in Mexico in the early 1970s. More than 10,000 confirmed cases were observed during 1972. Jorge Olarte, then at the Mexico Children's Hospital in Mexico City, characterized the strain of the bacterium *Salmonella typhi* responsible for the outbreak and demonstrated that it carried genes making it resistant to chloramphenicol (the drug of choice for this infection), ampicillin, streptomycin, and sulfonamides. Oddly enough, the same resistance pattern had been observed in Central America, but in *Shigella dysenteriae*—a different bacterial species. Presumably the genes encoding the antibiotic resistances were passed from the Central American *Shigella* to the Mexican *Salmonella* strain.

Lately, much attention has been focused on the resurgence of tuberculosis, a disease that had all but disappeared in industrialized nations thanks to the use of antibiotics.

In the past, these drugs were identified through massive drug-screening efforts. However, no new drugs for tuberculosis have emerged since the introduction of rifampin more than 30 years ago. Now, as multi-drug-resistant tuberculosis strains are isolated with increasing frequency, there has been renewed interest in identifying new therapeutic and prophylactic agents.

Drug-resistant tuberculosis is not new. By the late 1940s, only a few years after streptomycin proved to be the first effective anti-tuberculosis drug, resistant strains emerged. Shortly afterwards, clinicians realized that tuberculosis could easily gain resistance to a single drug and often to two. Three drugs, however, seemed invincible.

In keeping with that philosophy, the Centers for Disease Control and the Food and Drug Administration approved a combination drug containing rifampin, isoniazid, and pyrazinamide in the treatment of tuberculosis. Recently, however, not even a three-drug regime is sufficient to treat the newly emerged resistant strains. These are impervious to almost every available tuberculosis drug, and their emergence has resurrected the use of isolated tuberculosis wards in hospitals—a strategy dating back to the days before antibiotics were discovered. This would at least strike at the root of the problem in drug-resistant tuberculosis, which is patients failing to complete the full course of drug therapy.

Frequently patients start feeling well within 2 to 3 months of starting drug treatment, but it can take up to 18 months before all of the tuberculosis-causing microorganisms are killed. In the past, patients were kept in the hospital for the full course of treatment, so compliance was not a problem. But the move toward outpatient treatment and drug self-administration has fueled the rise in multiresistant organisms. Many people who fail to complete an adequate course of drug therapy relapse and require retreatment. Such circumstances create the conditions for the selection of drug-resistant organisms. For exam-

ple, in New York from 1982 to 1984, although 9.8 percent of *Mycobacterium tuberculosis* cells isolated from untreated patients were resistant to one or more drugs, 52 percent of isolates from relapsed patients were resistant.

Natural-Born Killers

People tend to think of antibiotics as a human invention when in truth they are perfectly natural. Ever since British biologist Alexander Fleming discovered in 1928 the antimicrobial activity of a substance released from the *Penicillium* fungus (the substance was called, aptly enough, penicillin), people have appreciated that organisms can manufacture powerful antibiotics. Antibiotics are, in fact, manufactured by the very classes of organisms they aim to destroy—bacteria and fungi. Following Fleming's discovery of penicillin, Selman Waksman from Rutgers University isolated streptomycin from the soil bacterium *Streptomyces griseus* in 1943.

Scientists are not entirely sure why organisms manufacture antibiotics, and the issue is a subject of some debate. It has often been stated that antibiotics are used by microorganisms as weapons against competing species. Such weapons, the theory goes, are especially useful to organisms that are colonizing a new niche. However, this to us seems somewhat inconsistent with certain features of antibiotics. One would expect that an organism in search of a new environment would not have the resources to make complex antibiotics. Therefore, one might expect antibiotics to be simple compounds, easily made from abundant materials. This, however, is not the case. Most antibiotics are complex and require a good deal of energy for their manufacture. Furthermore, antibiotics are produced by organisms in a stationary phase, which would seem incompatible with competition.

We, on the other hand, agree with Julian Davies of the University of British Columbia, who proposes that antibiotics are vestiges of ancient metabolic systems, dating back to some of the very first organisms on earth. Many antibiotics bind to cellular structures. Although today this

date	place	microorganism	disease manifestation	resistance	number sick
1960–1970	USA (St. Paul, Minn.)	*Neisseria gonorreae*	gonorrhea	penicillin	250,000–500,000
1963–1967	USA	*Salmonella typhi*	diarrhea	chloramphenicol, streptomycin, tetracycline, sufonamides	3,025 (43 deaths)
1968	USA	*Streptococci*	nosocomial (hospital-acquired) infections usually affecting respiratory and urinary tracts, but not exclusively	sulfonamides	1,020
1969–1970	USA	*Salmonella Klebsiella pneumoniae*	diarrhea	multiple drugs	19
1972–1973	Mexico	*Salmonella typhi*	typhoid fever	chloramphenicol, streptomycin, tetracycline, sulfonamides	10,000–15,000
1976–1979	Canada	*Salmonella typhi*	diarrhea	multiple drugs, including ampicillin, cephalothin, chloramphenicol kanamycin, nalidixic acid, nitrofurantoin, streptomycin, sulfonamide, tetracycline	416
1977	USA (Minnesota)	*Streptococci*	nosocomial infections	penicillin g	350
1977	USA (St. Paul)	*Staphylococcus aureus*	nosocomial infections	methicillin aminoglycosides	201
1983	USA (Virginia, Florida)	*Staphylococcus aureus*	nosocomial infections	methicillin	32
1986	USA	*Shigella sonnei*	nosocomial infections	ampicillin, tetracycline, sulfamethoxazole/trimethoprim	347
1988	Mexico	*Serratia marcescens*	nosocomial infections	ampicillin, carbenicillin, cefuroxim, netilimicin, gentamicin. tmp/smx, amikacin, phosphomicin	80
1989–1990	India	*Salmonella typhi*	diarrhea	chloramphenicol, ampicillin, tetracycline, streptomycin	37
1990–1992	USA	*Mycobacterium tuberculosis*	tuberculosis	multiple drugs	142
1991–1992	USA (New York)	*Enterococcus faecium*	diarrhea	vancomycin, ampicillin, gentamicin, streptomycin	7
1992	Philadelphia, New York, London	*Enterococcus faecalis Enterococcus faecium*	diarrhea	vancomycin	37
1992	Spain	*Staphylococcus aureus*	nosocomial infections	methicillin, aminoglycosides	NA
1992	USA	*Serratia marcescens*	nosocomial infections	multiple drugs	12 infants born prematurely

Outbreaks of bacteria resistant to more than one antibiotic have been documented since the early 1960s.

specific binding inhibits cellular activity, it could have once facilitated the synthesis of biological molecules such as peptides, or it could have stimulated other metabolic pathways. As biochemistry evolved, these binding molecules were likely replaced by enzymes, which proved more efficient metabolic facilitators. Nevertheless the ancient molecules persisted and now function as antibiotics.

Whatever their evolutionary significance, antibiotics have proved to be powerful weapons against other microorganisms. In general, antibiotics prevent the construction of crucial cellular components, prohibiting the target organisms from proliferating.

Some antibiotics, such as the beta-lactams, a class that includes the penicillins, and the cephalosporins, disrupt the construction of bacterial cell walls. Others, such as the tetracyclines, inhibit bacterial protein synthesis, and still others interfere with DNA or RNA synthesis. Antibiotics are medically useful not only because they kill off unwanted microorganisms, but also because they do not have similar effects on human cells, which are sufficiently different from bacterial cells to escape destruction. In

> **And research has shown that once an organism becomes resistant to certain antibiotics, it can potentially pass on the resistance to other members of its own species and even to different species.**

that they can be specifically targeted to particular microganisms, antibiotics have been called a "magic bullet." This specificity also differentiates antibiotics from antiseptics, which are generally toxic to a variety of cells, be they bacterial or human.

By the late 1950s, but particularly during the 1960s, an explosive search for natural and synthetic antibiotics was under way, eventually yielding about 100 new drugs. And as the costs of producing antibiotics dropped, these compounds were being used for just about any possible purpose: From

their appropriate use in therapy against infectious diseases, they soon went on to become supplements for animal feed, where they were supposed to be acting as "growth promoters." Thousands of pounds of antibiotics were released into the environment each year, killing off only the sensitive bacteria. The obvious result had been anticipated by Fleming himself: the emergence and dramatic increase in bacteria that were resistant to the effects of antibiotics.

Where does the resistance come from? Antibiotics are so potentially destructive to microorganisms that they threaten to disrupt the metabolism of the very same organisms that make them. As a result, many microbes develop mechanisms to protect themselves from their own antibiotics. It is this act of self-protection that makes organisms resistant to antibiotics. And research has shown that once an organism becomes resistant to certain antibiotics, it can potentially pass on the resistance to other members of its own species and even to different species.

Dodging the Magic Bullet

Antibiotic resistance, then, is rooted in the bacterial defense against its own harmful antimicrobials. The strategies by which microbes dodge the antibacterial bullet are as varied as the mechanisms of antibiotic activity.

In some cases, enzymes are manufactured by the host bacteria that dismantle the antibiotic molecules. For example, some bacteria acquire the ability to produce and secrete enzymes called β-lactamases, which degrade members of the β-lactam family of drugs before they even enter the cell. Enzymes can also be manufactured by bacteria to chemically modify antibiotics such as streptomycin, gentamicin, and amikacin so they lose their potency. In some cases, the bacteria can employ another strategy. Sometimes the bacteria modify the shape of the antibiotic's molecular target, so the antibiotic no longer recognizes it and is rendered inactive.

Some drugs, such as the tetracyclines, can successfully gain entry to the bacterium, but the resistant strains have acquired molecular pumps that

just pump the drugs right out of the bacteria.

Antibiotic resistance is dangerous when any organism becomes impervious to the effects of a drug. But what makes the current state of affairs even more dangerous is that many of the genes conferring resistance are found on plasmids, small circular pieces of DNA that can be easily exchanged between different bacteria of the same species, or even between individuals of different species. In fact, plasmid exchange between bacteria is probably the single most important mechanism for the spread of resistance genes. Resistance to the β-lactams, which include penicillin, and the aminoglycosides such as streptomycin, gentamicin, amikacin, and others is found on plasmids.

The variety of inactivating enzymes encoded on plasmids is incredible. George Miller and his team at the Schering-Plough Research Institute in New Jersey, for example, have isolated many of the 20 genes encoding enzymes that debilitate the aminoglycosides. He has also classified them based on their evolutionary path, discerning the origins and subsequent transfers of these genes among bacteria.

Recently, an additional mechanism allowing the spread of resistance genes has come to light. Some genes can move from one DNA molecule to others within the same cells. It has long been known that certain so-called mobile genetic elements existed in a variety of organisms, including bacteria. But scientists are just starting to appreciate how important these elements are in spreading and maintaining resistance genes within bacterial populations. For example, a single bacterial cell can contain many different plasmids, and genetic cassettes containing resistance genes can be exchanged between them. In this way a single plasmid may obtain several different resistance genes before it is passed on to another cell. Genetic mobility within bacteria augments opportunities for gene flux among different DNA molecules and between different cells within a microbial population.

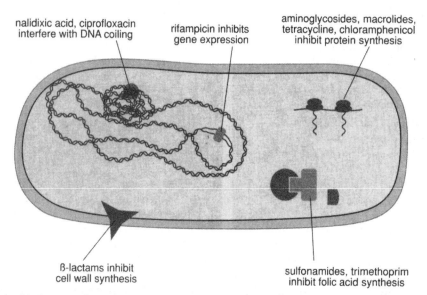

nalidixic acid, ciprofloxacin
interfere with DNA coiling

rifampicin inhibits
gene expression

aminoglycosides, macrolides,
tetracycline, chloramphenicol
inhibit protein synthesis

ß-lactams inhibit
cell wall synthesis

sulfonamides, trimethoprim
inhibit folic acid synthesis

Antibiotics generally work by disrupting the normal metabolic functions of the target bacteria.

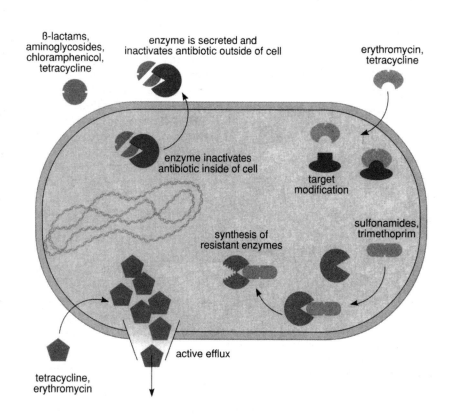

ß-lactams,
aminoglycosides,
chloramphenicol,
tetracycline

enzyme is secreted and
inactivates antibiotic outside of cell

erythromycin,
tetracycline

enzyme inactivates
antibiotic inside of cell

target
modification

synthesis of
resistant enzymes

sulfonamides,
trimethoprim

active efflux

tetracycline,
erythromycin

Resistant bacteria develop a variety of mechanisms to dodge the antibiotic bullet. Some bacteria unleash enzymes that can dismantle antibiotics such as the β-lactams, aminoglycosides, chloramphenicol, and tetracycline before they even enter the bacterial cell. Sometimes, bacteria develop pumps that shunt antibiotics such as tetracycline and erythromycin out of the cell. Another tactic is to alter the shape of a bacterial molecule to which the antibiotics erythromycin or tetracycline must bind.

Where do resistance genes come from initially? Most of them are identical or very similar to those found in antibiotic-producing organisms. It is possible that these genes traveled from the microbes in which they first arose, to the ones responsible for infections in people and animals. In 1973 Julian Davies and Raoul Benveniste (both then at the University of Wisconsin) first noticed the similar biochemical mechanisms of resistance between the aminoglycoside-producing bacteria and common pathogenic microbes. More recently Davies has found DNA sequences containing antibiotic-resistance genes in commercial antibiotic preparations. This DNA may be taken up by pathogenic bacteria even during the course of antibiotic therapy.

A number of bacterial responses to environmental stress result in the increased resistance of bacteria to antimicrobial drugs. Unlike the previously described mechanisms of resistance, the genes encoding these resistance mechanisms reside not on plasmids, but on the bacteria's main chromosome. Genes that encode multiple antibiotic resistance, called *mar* genes, described by Stuart Levy of Tufts University, are turned on in the presence of at least two different antibiotics, tetracycline and chloramphenicol, as well as salicyclate, which is the active ingredient of aspirin. The drugs act in this case as an environmental stressor. The presence of these stressors leads to a temporary bacterial resistance to many other drugs.

Many drugs enter a bacterium via molecular pores on the surface of the bacterial target. The *mar* gene products initiate a series of events that ultimately block the expression of pore genes and thus inhibit pore formation. When this happens, the outer membrane becomes less permeable to drugs, which then cannot reach their targets within the cells. This mechanism confers resistance to several unrelated types of antibiotics—all those that rely on the pores to gain entry to the bacterium. This form of resistance cannot last forever, however, as the membrane becomes less permeable not only to unwanted drugs, but also to much-needed nutrients. Pore inhibition lasts only as long as the bacterium senses the chemical stress, in this case the presence of the antibiotics. When the antibiotics are gone, the pores reform, and the bacterial functions return to normal.

Another regulated system, which overlaps with *mar*, is the response to superoxide, which are produced by human immune cells to destroy bacteria. The bacterial response to superoxide is governed by the bacterial soxRS genes, described by Bruce Demple at the Harvard School of Public Health. The *soxRS* genes, like the *mar* genes, reduce the expression of membrane pores and make the bacteria resistant to both antibiotic drugs and the immune system's superoxide. Bacterial resistance to immunological superoxides renders the bacteria more virulent since the immune system becomes ineffective at disabling it.

The resistance genes involved in both the *mar* and *soxRS* systems are found on the main bacterial chromosome. Genes in this category are generally involved in cell regulation. Sometimes genes encoded on plasmids can become irreversibly incorporated into the chromosome, but this is rare. Chromosomal mutations usually reduce the ability of the bacterium to survive in environments free of antibiotics and may also render the bacterium less virulent. Finally, chromosomal resistance genes, unlike resistance genes on plasmids, are unlikely to be transferred between different bacteria.

Growing Together

Interactions between bacteria are facilitated by the way bacterial colonies are formed, placing individual bacteria, sometimes of different species, in close proximity. Bacterial colony formation may also help to explain some forms of resistance and should be taken into account by researchers developing antibiotics.

Antibiotics were and still are developed on the basis of test-tube or *in vitro* experiments. Clinical tests, or antibiograms, to determine the antibiotic sensitivity of bacteria found at

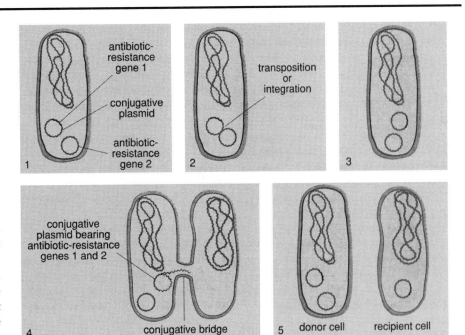

Antibiotic resistance in bacteria is especially dangerous because the genes encoding resistance can be transferred between individuals of the same species or between individuals of different and distantly related bacterial species. Resistance genes are most often found on small, circular pieces of DNA called plasmids, which are independent from the main chromosome. A single bacterium can contain many different plasmids, each with a different set of resistance genes on it. In addition, resistance genes can be transferred from one plasmid to another. In this way, a single plasmid can collect a number of different resistance genes. Such is the case in panel 2, where a resistance gene is copied from one plasmid onto another that already contains a different resistance gene. This multiply resistant plasmid is transferred to another bacterial cell in panel 4 by conjugation, which is one of the major mechanisms by which DNA is transmitted to different cells. The formerly antibiotic-sensitive recipient cell in panel 5 now contains the multiply resistant plasmid. The recipient cell is now antibiotic resistant.

sites of infection, are also carried out on test-tube-grown bacteria. In both cases, bacteria are grown "planktonically," that is, in homogeneous suspensions of liquid medium. This condition may not reflect the way in which bacteria grow naturally when causing an infection and, in particular, may result in a substantially different susceptibility to antibiotics.

Bacteria in aquatic environments, both natural and artificial, and in infected tissues have an extraordinary tendency to interact with surfaces to form associations called biofilms. These cooperative consortia have been studied by J. William Costerton and his group at the Center for Biofilm Engineering in Montana. The bacterial cell produces a variety of complex sugars to form a coat named the glycocalyx, which assists in firmly gluing the cell to inert surfaces. Cell division

takes place inside the glycocalyx matrix and results in the formation of microcolonies. Continuous reproduction and the attachment of bacterial cells of different species result in a heterospecific biofilm.

The application of confocal scanning laser microscopy to the study of biofilms revealed an amazing architecture. Highly permeable water channels exist to permit the penetration of large molecules in a primitive "circulatory system" that delivers nutrients from the environment to the microclimate niche and removes the metabolic waste products at the same time. Bacteria within a mature multispecies biofilm live in a special microniche where nutrients are provided by neighboring cells and by diffusion. In this way, microcolonies of cells capable of primary production of nutrients are often surrounded by microorganisms with larger nutrient requirements. The physiology of biofilm cells is very complex and different from that of cells grown planktonically. The physiological status of attached cells can vary depending on the location of each individual cell within the multiple layers in the biofilm.

Bacteria obtain a number of advantages by living in biofilms. Cells are protected from environmental stresses such as heat, ultraviolet radiation, and viruses. They are also protected from antibacterial agents. The resistance of biofilms to antibacterials can have serious medical consequences, since biofilms can form on catheters, implants, dental units, silicon surfaces, contact lenses, endotracheal tubes, and so on. Antibiotics cannot eliminate these biofilms. The only way to arrest the infection is to replace the affected device.

The structure of biofilms changes the susceptibility of resident bacteria to antibacterial agents. Bacterial cells grow embedded in a thick layer that constitutes a solute phase distinct from the bulk fluid phase of the system. Biofilms generate a sheltered, encapsulated community of cells in which environmental stresses are greatly reduced. Changes in the surrounding medium cause alterations in the susceptibility of bacteria to antibiotics. The resistance of *Pseudomonas aeruginosa*, for example, to polymyxin B and aminoglycosides increases under conditions of magnesium depletion (a likely situation within a biofilm). In addition, it has been shown that biofilm cells produce 32 times more β-lactamase enzymes, which destroy the β-lactam family of antibiotics, than do cells of the same strain grown planktonically.

The formation of biofilms may be the heart of some kinds of antibiotic resistance. It is possible, for example, that the glycocalyx forms an impermeable barrier to antibiotics, so individuals buried within the biofilm are never exposed to the drugs. Living in a biofilm may cause the activation of genes associated with a sedentary existence. Coincidentally, these genes may also affect the susceptibility of the bacteria to drugs.

It is also possible that the limited availability of key nutrients within the biofilm slows down the growth rate. As a result, some of the bacteria, particularly at the base of the biofilm, may be in a dormant state. Some kinds of antibiotics exert their influence only on actively growing cells, so dormant cells would resist those antibiotics.

Planktonic and biofilm cells are known to coexist at sites of infection. When these cells are exposed to antibiotics, planktonic cells and those at the surface of the biofilm are quickly affected. The excess antibiotic molecules that have entered the cells and

> It is important to understand that the outcome of any, or perhaps all, strategies is unpredictable, as science is playing a catch-up game with bacteria that are continually evolving.

that are not engaged in cell inactivation are probably destroyed by antibiotic-degrading enzymes or are involved nonspecifically with other cellular components. The rest of the antibiotic molecules are trapped by glycocalyx, which is negatively charged and is known to function as an ion-exchange resin (like those used to purify water and other compounds), capable of binding a very large number of antibiotic molecules.

To overcome resistance brought about by biofilms, new drug strategies are needed. Combinations of antibiotics have been used successfully against some biofilms. For prosthetic devices and biocompatible materials, however, the future is uncertain. For example, the bacteria *Proteus* and *Pseudomonas* bind effectively to orthopedic devices coated with gentamicin that were specifically designed to prevent biofilm formation, as shown by Chung-Che Chang and Katharine Merritt at Case Western Reserve University.

Facing Resistance

Taking into account a more realistic model for bacterial growth in biofilms may go a long way toward improving drug development and testing strategies. But of course other approaches must be taken as well. It is important to understand that the outcome of any, or perhaps all, strategies is unpredictable, as science is playing a catch-up game with bacteria that are continually evolving.

antibiotic	resistance mechanism
β-lactams aminoglycosides chloramphenicol erythromycin tetracycline	chemically modified by enzymes
tetracycline erythromycin	actively removed from cell
erythromycin	enzymatic modification of target
β-lactams fusidic acid	proteins bind to and sequester antibiotics within target cell
sulfonamides trimethoprim	synthesis of enzymes insensitive to the action of the drug

Resistances to a number of antibiotics are transferable in bacteria.

In our laboratory, we have been looking for ways either to eliminate resistance-encoding plasmids from their bacterial hosts or to prevent the expression of resistance genes encoded on them. There are several reported chemical agents capable of doing that, but most of them are very toxic. Some agents eliminate plasmids at concentrations that kill the bacteria. If this happens, a resistance mechanism will soon arise because of the strong selective pressure.

So far, the most promising agent has been ascorbic acid, better known as vitamin C, which has effectively eliminated resistance plasmids from *Staphyloccoccus aureus*. Ascorbic acid may not turn out to be the ultimate drug for plasmid elimination, since very high concentrations are needed to obtain the effect, but it may lead the way to similar, more effective compounds. Another novel approach to eliminating resistance is taken by Jack Heinemann at the Rocky Mountain Laboratories in Montana. He has proposed the possibility of halting the spread of resistance genes by inhibiting gene transfers between bacteria.

From the point of view of the pharmaceutical industry, the simplest way to face the threat of antibiotic resistance is to chemically modify existing drugs. Modified compounds, also called semisynthetic antibiotics, usually have the same or slightly improved pharmacological properties as existing drugs, but are resistant to the inactivating enzymes. Isepamicin, for example, is in the aminoglycoside family of antibiotics, but is insensitive to most of the aminoglycoside-inactivating enzymes found among resistant bacteria. The so-called third-generation cephalosporins include drugs resistant to some β-lactamases. However, if modifying the drug molecule is relatively easy, it seems just as easy for the bacteria to evolve a new deactivating enzyme in response.

Another possible approach to circumvent resistance is the development of agents that inhibit the resistance mechanisms themselves, restoring bacterial susceptibility to old drugs. Several compounds that inhibit the β-lactamase enzymes are already

in clinical use. Other compounds that diminish tetracycline resistance are currently being studied. Stuart Levy and his colleagues at Tufts are trying to identify molecules that are structurally similar to tetracycline, which can bind to and clog up the molecular pumps that shunt tetracycline out of the cell in resistant bacteria. That way, actual tetracycline molecules can be used again for as long as the pumps remain gummed up.

In addition to rehabilitating old antibiotics, pharmaceutical scientists are trying to develop completely new antibiotics. Some groups are screening numerous potential antibiotic sources and then identifying and isolating the active molecules. Their search has

produced squalamine, a steroid isolated from sharks, as well as cryptdin and cecropin, isolated from the mammalian intestine, and ranalexin, which was derived from frog skin.

Investigators are also going back to the original source of antibiotics— microorganisms. They are screening organisms such as soil bacteria to see whether any new and useful antibiotics have appeared. For example, such a search led research teams headed by George Miller at Schering-Plough to find everninomycin, which may serve as the prototype for an entirely new family of antibiotic drugs, for which no resistances have so far been discovered. But the potential for resistance is one possible dis-

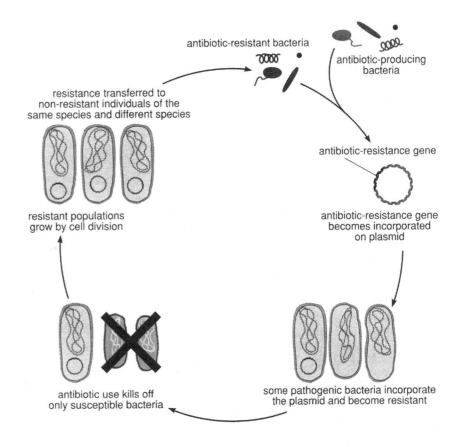

Cycle of resistance must be broken if antibiotics are once again to be useful. The cycle begins with soil bacteria that produce antibiotics. These organisms also produce resistance mechanisms that protect the bacteria from their own antibiotics. The genes encoding resistance make their way onto plasmids, via a series of gene exchanges and integrations. These plasmids become incorporated into pathogenic bacteria, rendering them resistant to antibiotics. When people take drugs, they kill only the susceptible bacteria. The resistant bacteria remain behind and multiply, producing resistant offspring. Resistant bacteria can exchange resistance genes with other bacteria, rendering antibiotic-sensitive organisms resistant, and singly resistant organisms multiply resistant.

advantage in microbial sources for antibiotics. The organisms producing these antibiotics probably already have resistance mechanisms against the drugs. If the genes encoding these mechanisms are located on plasmids or any other transferable genetic molecule, the resistance can spread to disease-causing bacteria, which would set the cycle in motion again.

Another sophisticated way to direct the search for new antibiotics is rational drug design, which uses information about the structures of molecules to create or modify drugs. Researchers at Hoffmann-La Roche in Switzerland have already rationally designed inhibitors of antibiotic-modifying enzymes.

The search for and development of new drugs by the pharmaceutical industry will go a long way toward conquering the growing microbial resistances to available antibiotics. But there is much more to be done than merely generating new antibiotics—the pace of which cannot keep up with the microbial resistance responses.

A significant change of attitude must also be encouraged among physicians, industry, and the public. Antibiotics must be seen as a valuable resource that should be used carefully and only when really needed. The judicious use of antibiotics and the dangers of bacterial resistance must be taught worldwide in the early years of medical training, and it must continue to be taught to medical practitioners throughout the course of their careers. The efforts undertaken by the Alliance for the Prudent Use of Antibiotics led by Stuart Levy must receive much stronger support. Industries must stop pushing for the nonclinical use of antibiotics, which accounts for most of the sales of several drugs. The pharmaceutical industry must realize that it will benefit from the rational use of antibiotics and should provide financial support for these efforts.

Finally, a worldwide campaign to eliminate over-the-counter preparations of antibiotics, especially in countries where all antibiotics are sold over-the-counter, should be promoted. The spread of antibiotic-resistance genes is not restricted by political geography, and self-administered antibiotics may account for a significant fraction of antibiotic misuse.

The history of antibiotics reminded our student Raul Borbolla of the Greek myth of Sisyphus, the king of Corinth who, as a punishment for his hubris, was condemned by the gods to push a boulder up a mountain, only to have the boulder roll to the bottom, from which Sisyphus had to start pushing again. Again the boulder would roll to the bottom, and the cycle was repeated into perpetuity. The rational and controlled use of antibiotics may prevent medicine from facing Sisyphus's fate.

Acknowledgments

We thank Marina Chicurel and Julian Davies for their kind and useful comments, and Raul Borbolla, who assisted with the bibliographic research. This article is dedicated to the memory of Dr. Julian Villarreal, mentor and friend.

Bibliography

Amábile-Cuevas, C. F. 1993. *Origin, Evolution and Spread of Antibiotic Resistance Genes.* Austin, Texas: R. G. Landes Co.

Amábile-Cuevas, C. F. and M. E. Chicurel. 1993. Horizontal gene transfer. *American Scientist* 81:332-341.

Amábile-Cuevas, C. F. and M. E. Chicurel. 1992. Bacterial plasmids and gene flux. *Cell* 70:189-199.

Anwar, H., J. L. Strap and J. W. Costerton. 1992. Establishment of aging biofilms: Possible mechanism of bacterial resistance to antimicrobial therapy. *Antimicrobial Agents and Chemotherapy* 36:1347-1351.

Ariza, R R., S. P. Cohen, N. Bachhawat, S. B. Levy and B. Demple. 1994. Repressor mutations in the *marRAB* operon that activate oxidative stress genes and multiple antibiotic resistance in *Escherichia coli. Journal of Bacteriology* 176:143-148.

Brown, M. R. W. and P. Gilbert. 1993. Sensitivity of biofilms to antimicrobial agents. *Journal of Applied Bacteriology* 74:87S-97S.

Costerton, J. W., Z. Lewdnowski, D. DeBeer, D. Caldwell, D. Korber and G. James. 1994. Biofilms, the customized microniche. *Journal of Bacteriology* 176:2137-2142.

Courvalin, P. 1994. Transfer of antibiotic resistance genes between gram-positive and gram negative bacteria. *Antimicrobial Agents and Chemotherapy* 38:1447-1451.

Davies, J. 1990. What are antibiotics? Archaic functions for modern activities. *Molecular Microbiology* 4:1227-1232.

Davies, J. 1993. Grandeur et decadence des antibiotiques. *Le Recherche* 24:1354-1361.

Davies, J. 1994. Inactivation of antibiotics and the dissemination of resistance genes. *Science* 264:375-382.

Heinemann, J. A. 1993. Transfer of antibiotic resistances: A novel target for intervention. *APUA Newsletter* 11:1-6.

Jarlier, V. and H. Nikaido. 1994. Mycobacterial cell wall: Structure and role in natural resistance to antibiotics. *FEMS Microbiology Letters* 123:11-18.

Levy, S. B. 1992. *The Antibiotic Paradox.* New York: Plenum Press.

Sebald, M. 1994. Genetic basis for antibiotic resistance in anaerobes. *Clinical Infectious Diseases* 18 (supplement 4):S297-S304.

Shaw, K. J., P. N. Rather, R. S. Hare and G. H. Miller. 1993. Molecular genetics of aminoglycoside resistance genes and familial relationships of the aminoglycoside-modifying enzymes. *Microbiological Reviews* 57:138-163.

Snider, D. E. and J. R. La Montagne. 1994. The neglected global tuberculosis problem: A report of the 1992 World Congress on Tuberculosis. *Journal of Infectious Diseases* 169:1189-1196.

Tomasz, A. 1994. Multiple-antibiotic-resistant pathogenic bacteria. *New England Journal of Medicine* 330:1247-1251.

Weis, S. E., P. C. Slocum, F. X. Blais, B. King, M. Nunn, B. Matney, E. Gómez and B. H. Foresman. 1994. The effect of directly observed therapy on the rates of drug resistance and relapse in tuberculosis. *New England Journal of Medicine* 330:1179-1184.

THOUGHT EVOKERS

• Describe the mechanisms responsible for keeping antibiotics from destroying disease-causing bacteria.

• Historically, what causes have been associated with outbreaks of infectious diseases?

• Although many people believe that antibiotics are a human invention, they are basically of natural origin. Explain why this statement is true.

7

CLIMATE AND THE RISE OF MAN

The history of human evolution holds sobering lessons for those gathering at this week's Earth Summit

By William F. Allman with Betsy Wagner

When global warming finally came, it struck with a vengeance. In some regions, temperatures rose several degrees in less than a century. Sea levels shot up nearly 400 feet, flooding coastal settlements and forcing people to migrate inland. Deserts spread throughout the world as vegetation shifted drastically in North America, Europe, and Asia. After driving many of the animals around them to near extinction, people were forced to abandon their old way of life for a radically new survival strategy that resulted in widespread starvation and disease. The adaptation was farming: the global-warming crisis that gave rise to it happened more than 10,000 years ago.

As environmentalists convene in Rio de Janeiro this week to ponder the global climate of the future, earth scientists are in the midst of a revolution in understanding how climate has changed in the past—and how those changes have transformed human existence. Researchers have begun to piece together an illuminating picture of the powerful geological and astronomical forces that have conspired to change the planet's environment from hot to cold, wet to dry, and back again over a time period stretching back hundreds of millions of years.

Most important, scientists are beginning to realize that the gyrations of this climatic dance have had a major impact on the evolution of the human species. New research now suggests that climate shifts have played a key role in nearly every significant turning point in human evolution: from the dawn of primates some 65 million years ago to human ancestors rising up to walk on two legs, from the prodigious expansion of the human brain to the rise of agriculture. Indeed, the human saga has not been merely touched by global climate change, some scientists argue, it has in some instances been *driven* by it.

The new research has profound implications for the environmental summit in Rio. Among other things, the findings demonstrate that dramatic climate change is nothing new for planet Earth. The benign global environment that has existed over the past 10,000 years—during which agriculture, writing, cities and most other features of civilization appeared—is a mere blip in a much larger pattern of widely varying climate over the eons. In fact, the pattern of climate change in the past reveals that Earth's climate will almost certainly go through dramatic changes in the future—even without the influence of human activity.

At the same time, the research provides little comfort for those who would like to believe the Earth is a self-regulating machine that can unfailingly absorb the impact of any human activity. Over Earth's history, tiny alterations in the positions of the continents, the flow of air currents, and other influences on the world's weather sometimes cascaded into huge changes in global climate. If the study of prehistory is any guide, a large shift in climate is likely to bring a fundamental change in the nature of human life.

If not for a dramatic climate shift some 65 million years ago, most of the animals on Earth today—including humans—would probably not even be here. Scientists have long suspected that a giant meteor collided with the Earth at that point in time, sending huge clouds of climate-altering dust into the atmosphere. The recent discovery in the Caribbean of tiny nuggets of glass whose chemical makeup suggests that they were formed in the heat of such a cosmic collision lends new support to the theory.

Breadfruit in Greenland. Scientists find evidence that in the heyday of the dinosaurs, 100 million years ago, the world was 10 to 14 degrees warmer than it is today. Breadfruit trees grew in what is now Greenland and dinosaurs wandered an ice-free Antarctica. In the wake of the meteor's impact, dinosaurs vanished in massive numbers, leaving the world wide open for colonization by mammals, including a small, shrewlike creature that was the ancient ancestor of humans.

Most shifts in Earth's climate have not been so sudden or dramatic. But even slowly changing environments have had an enormous influence on the evolution of the human species. After the demise of the dinosaurs, for instance, the Earth continued to grow cooler for tens of millions of years. The cooling resulted from the slow absorption into the Earth of atmospheric carbon dioxide through the weathering of rock, suggests Yale University's Robert Berner, who recently used computer modeling to show how carbon dioxide levels in the atmosphere have changed over the past 600 million years. Because carbon dioxide traps heat to create the so-called greenhouse effect, over time the reduction in CO_2 in the Earth's atmosphere made the global temperature drop several degrees.

This gradual cooling helped set the stage for a crucial phase of human evolution: the beginning of upright walking. Ever since Darwin, anthropologists have speculated that our ancestors rose onto two legs in order to free their hands for some uniquely human activity, such as making tools. But Peter Rodman and Henry McHenry of the University of California at Davis argue instead that the first bipeds were trying to maintain their apelike lifestyle amid environmental change.

New evidence indicates that the Earth's cooling climate caused the dense forests that blanketed Africa to break up into clumps of trees separated by open patches of grassy savanna. Geochemist Thure Cerling of the University of Utah has pioneered a technique that measures tiny amounts of carbon left behind in the soil as plants died. Grasses leave a different chemical "signature" in the soil than do trees. By analyzing the composition of ancient soils, Cerling has discovered that grasses began to appear and spread in Africa about 7 million years ago.

The African environment's gradual shift to a patchwork of forests and plains made walking on two legs an ideal way for human ancestors to move about, Rodman and McHenry contend. Analyzing how much energy various animals use as they walk, the researchers calculated that walking upright is a more efficient way to travel long distances than is walking on the feet and knuckles in the manner of chimps and gorillas. They conclude that our ancestors rose up on two legs not in order to make tools—which do not appear in the archaeological record until more than a million years later—but because walking was simply an easier way for the tree-loving creatures to travel from one clump of forest to another.

One of Earth's most dramatic climatic shifts coincided almost exactly with the most important turn in human evolution: the emergence of the first large-brained primate. An analysis of the chemical makeup of seashells taken from the deep sea floor suggests that about 2.5 million years ago, Earth's temperature dipped suddenly and dramatically. The rapid cooling appears to have been global in scope: For the first time in Earth's history, an icecap formed at the North Pole, and the continents became drier and dustier.

200 Million Years Ago

Average Temperature: 70°F

Turning Point: The appearance of the first mammals

All the land masses that now form the continents were once part of a single supercontinent that scientists call Pangea. Thrown together by massive geological forces, the colliding land masses pushed up enormous mountain ranges, remnants of which exist today as the Appalachians in the United States and the Urals in Russia. Pangean environments ranged from lush tropics to torrid deserts, and in some areas the temperature varied as much as 130 degrees. Across this land roamed huge reptiles, the ancestors of the dinosaurs, and a tiny, badgerlike creature that is the distant ancestor of nearly all mammals—including humans.

65 Million Years Ago

Average Temperature: 73°F

Turning Point: Dinosaurs become extinct; first primates appear.

Ripped apart by upwelling lava, Pangea broke up into land masses roughly the shape of the modern continents, which continued to wander toward the positions they now occupy. For 100 million years, the world was dominated by dinosaurs. A host of evidence suggests that about 65 million years ago, the Earth was rammed by a huge meteor. The collision sent clouds of dust into the atmosphere and triggered a drastic climate change. The dinosaurs quickly became extinct, leaving the world wide open for mammals to thrive in. One of those mammals was the first primate, the ancestor to all modern monkeys, apes, and humans.

Basic Data: Christopher Scotese/University of Texas at Arlington; "Paleoclimatology"/Thomas J. Crowley and Gerald R. North.

Himalayan highs. One cause of the cold snap may have been the skyward thrusting of the Himalaya Mountains which began rising when the Indian subcontinent collided with Asia some 40 million years before. Using a computer simulation of the flow of air currents around the world, William Ruddiman of Columbia University's Lamont-Doherty Geological Observatory and John Kutzback of the University of Wisconsin found that by 2.5 million years ago, the Himalayas' peaks towered high enough to alter the flow of major air currents in the Northern Hemisphere, significantly shifting weather patterns.

The drying of Africa opened the way for new kinds of life to emerge. Yale University paleontologist Elisabeth Vrba points out that African fossils from several species of animals indicate a dramatic change at precisely the same time as the climate shift. Many kinds of antelopes that made their home in woody environments, for example, suddenly disappeared about 2.5 million years ago and were replaced by new types of antelopes that lived in more open, grassy landscapes. Similarly, various rodent species that thrived in wet environments became extinct and were replaced by new kinds of rodents that thrived in drier climates.

Vrba argues that this "pulse" of climate change 2.5 million years ago also created conditions that caused ancient human forebears to undergo a burst of biological innovation. "Our ancestors behaved like perfect mammals," says Vrba. The fossil record shows that at about that time, the species of human ancestor that had existed for more than a million years disappeared, and three or possibly four new upright primates emerged. One of these, a creature that chipped stones to make tools, is the first to be called homo.

With the onset of cooler temperatures some 2.5 million years ago, Earth suddenly entered a whole new pattern of cyclical climate change that acted as an "evolutionary pump," causing the brains of ancient humans to grow larger and larger. Before that landmark, large shifts in Earth's climate typically were the result of geologic processes that took place over millions of years. After 2.5 million years ago, however, the global climate pattern suddenly became much more sensitive to minor perturbations in Earth's orbit and in the tilt of Earth's axis as it circled the sun.

Glacial cycles. These astronomical cycles have caused Earth's climate to fluctuate rapidly between long periods of cold lasting 50,000 to 80,000 years and shorter periods of warmth lasting, on average, about 10,000 years. During the colder periods, the Earth has been locked in ice, with glaciers extending far into Europe,

5 Million Years Ago

Average Temperature: 60°F

Turning Point: Chimpanzee and human lineages split from their common ancestor; human ancestors begin to walk on two legs.

As the continents continued to settle into their present positions, a geologic shift cut the Mediterranean Sea off from the Atlantic. Over a period of a thousand years the sea evaporated, leaving barren flats of salt. When sea levels rose again, water flowed back into the Mediterranean basin with a force hundreds of times greater than Niagara Falls. The cycle repeated itself again and again. Meanwhile, the Himalayas pushed higher, changing the flow of atmospheric winds and weather. The world cooled, and the great forests of Africa were broken up by patches of grassy plain. At about this time the family of primates split, one branch leading to modern apes and the other to humans. Fossilized footprints show that by 3.75 million years ago, these ancestors were walking upright.

2.5 Million Years Ago

Average Temperature: 55°F

Turning Point: Appearance of the first humans

The Himalaya Mountains pushed even higher, disturbing the flow of the atmospheric currents in the Northern Hemisphere. For the first time in Earth's history, a northern polar icecap appeared. As the Earth grew cooler, the vast tracks of dense woods disappeared from many parts of Africa; fossil evidence from the period shows that many species of forest-dwelling antelopes vanished, and many new species of antelopes that made their home in patchier, more open environments suddenly appeared. The line of upright primates splintered into several new species, including *Homo habilis*, a toolmaking creature whose brain was larger than that of any previous primate. It marked the beginning of the human lineage. At this point, Earth's climate shifted into a new pattern of cyclical change that has continued to the present: Tiny, periodic fluctuations in the Earth's tilt and the path of its orbit around the sun cause repeating cycles of long, cold periods—lasting 50,000 to 80,000 years—and episodes of warmth lasting about 10,000 years. The Earth today is near the end of one of these brief warm spells.

Asia, and North America, sea levels falling dramatically, and the land masses near the equator becoming cooler and drier. In the warmer, "interglacial" periods—the Earth is in one today—the glaciers retreat and the climate becomes more temperate. Earth has undergone at least 15 such cycles since about 1.6 million years

> **Though the rise of agriculture is sometimes depicted as an "invention" that was a pivotal step forward in human progress, it was more likely a last-ditch effort for survival in a rapidly deteriorating environment . . .**

ago, and if this climate pattern holds, another chilly plunge into glaciation is due to arrive within 2,000 years or so.

Some scientists suggest that these rapid shifts between warm and cold

were the driving force behind the human species' ballooning brain. The periods of warm, wet climate caused a population boom among our ancestors, argues neuroscientist William Calvin of the University of Washington. Then a new bout of cold weather would arrive, bringing harsh conditions that would winnow out all but the most intelligent creatures. When the "boom times" came again, these brainier human ancestors would grow in number, increasing their percentage in the overall population. The last time the Earth was as warm as it is today began about 125,000 years ago—a period that lasted roughly 10,000 years before another cold snap set in. Intriguingly, the first fossils of Homo sapiens—anatomically modern humans—come from about this time.

Even after modern humans had established themselves throughout much of the world some 30,000 years ago, the waxing and waning of Earth's

glaciers continued to have a major impact on how they conducted their lives. As the world was plunged into a harsher climate, human societies appear to have responded with increasing complexity. Some researchers believe that the magnificent artwork that graces the cave walls of Europe, for instance, was part of a spiritual or religious reaction to the difficult conditions caused when the huge glaciers reached their southernmost point 18,000 years ago.

Though the rise of agriculture is sometimes depicted as an "invention" that was a pivotal step forward in human progress, it was more likely a last-ditch effort for survival in a rapidly deteriorating environment caused when the glaciers began to recede again 15,000 years ago, argues anthropologist Ofer Bar-Yosef of Harvard University. It was then that the human race got its first real taste of global warming. Sea levels rose nearly 400

125,000 Years Ago

Average Temperature: 59°F

Turning Point: Anatomically modern humans appear.

The last time Earth's climate was as warm as it is today was 125,000 years ago, when the glaciers that had waxed and waned more than a dozen times receded once again. By this time, the human ancestor *Homo habilis* had given way to another species of early human known as *Homo erectus*. These creatures were the first human ancestors to leave Africa, migrating out from the continent about 1 million years ago and spreading throughout much of the Old World. When the warm spell ended about 115,000 years ago, the Earth's climate plunged into another intense bout of cold and glaciation. Intriguingly, it is around the time of this last climate shift that the first modern-looking humans—*Homo sapiens*—began to appear in Africa and the Near East. Over the next 80,000 years, modern humans would migrate throughout the Earth, from the frozen tundra of the Arctic to the tip of South America.

18,000 Years Ago

Average Temperature: 50°F

Turning Point: Humans live through the coldest climate in Earth's history.

At the peak of Earth's last period of intense glaciation, mile-thick glaciers reached down to what is now Chicago and covered most of New England. Glaciers also blanketed much of Europe, and those areas that were not locked in ice were covered with frozen tundra. The Cro-Magnon humans of Europe were well into the explosion of culture that would yield thousands of art objects—including the magnificent paintings that grace the walls of caves in southern France and Spain. Bone needles from the time suggest that humans were making warm, close-fitting clothing. Bows and arrows from soon after indicate that they had intensified their hunting-and-gathering way of life.

feet in some areas, cutting off Alaska from Siberia and New Guinea from Australia. The tundra that had covered much of Europe, Asia, and North America was replaced by forest and grass, leaving the climate much like today's.

The changes wrought by global warming were acutely felt in the Levant, an area in the Middle East now occupied by Israel, Syria, Lebanon, and Jordan. Some 15,000 years ago, the Levant was a lush hill country filled with a bounty of plants and animals, making it easy for humans to hunt and gather the food they needed to thrive. But as the world became warmer, the region experienced repeated droughts, and vegetation shifted northward. One group of hunter-gatherers who lived in the Levant, the Natufians, became increasingly hemmed in by the mountains to the north and spreading deserts to the south and east. According to Bar-Yosef, the Natufians responded to their ever diminishing range by abandoning their wandering ways and settling down, although they continued to hunt and gather food.

Bar-Yosef's theory is buttressed by new research by Daniel Lieberman of Harvard University, who examined the patterns of growth in the fossilized teeth of gazelles at sites in the Levant. Lieberman found that sites made by the Natufians' immediate predecessors contain the remains of gazelles that were killed in only a single season, such as the spring or fall. At Natufian sites, however, the gazelle remains are from animals that were killed year-round, indicating that the Natufians were living at a permanent camp. It is at about this time also that fossils of the "house" mouse began to appear at Natufian sites.

A time to sow. Just as the Natufians were adjusting to a new sedentary life in their drier environment, however, they suddenly were hit by another bout of climate change, this time to much cooler temperatures. Scientists are still unsure why, but about 11,000 years ago the Earth's climate returned to a glacial climate for nearly 1,000 years. The environment in the Levant quickly grew cooler, changing the plant and animal life, and the Natufians found themselves with a tough choice, says Bar-Yosef: They had to either begin moving again or find a new source of food. The Natufians started cultivating grains to survive.

The Natufians' shift to agriculture required no grand technological breakthroughs, argues Bar-Yosef. Despite their hunting-and-gathering lifestyle, the ancient people of the Levant appear to have been familiar for millenniums with cereals and grains: Scientists recently unearthed grains of wild barley at a newly discovered archaeological site in Israel that dates back 19,000 years.

11,000 Years Ago

Average Temperature: 59°F

Turning Point: Humans begin to depend on agriculture.

The glaciers receded as an intense bout of "global warming" began. The lush, nearly Eden-like conditions that had existed in the Middle East suddenly shifted northward, leaving arid land behind. A group of people known as the Natufians became concentrated in a small, fertile area bounded by mountains, sea, and desert, and they ceased their nomadic ways. A sudden climate shift led to a cold spell that lasted for 1,000 years, forcing the Natufians to turn to farming for survival. When the climate finally warmed up again, a population explosion among the Natufians further locked them and their descendants into an agricultural life.

The rapid swings of climate that occurred between 13,000 and 10,000 years ago—and the shift to agriculture that resulted from them—must have created a wrenching change in how people lived. After all, people had been hunting and gathering for as long as modern humans had walked the earth. Research by Mark Cohen, an anthropologist at the State University of New York at Plattsburgh, suggests that the first farmers suffered malnutrition, vitamin deficiencies, and a host of new diseases that arose because of their sedentary ways. Social life changed, too, since people could no longer simply walk away from disputes and so had to devise new social strategies for resolving conflicts. The increasing reliance on stored goods meant that for the first time people had possessions that could be stolen, creating the need for defensive walls and militia.

Out of Eden. One of the epic "origin" tales to come out of the region—the biblical book of Genesis—speaks of humans being expelled from a garden of plenty and being punished by having to "eat the herb of the field" and "in the sweat of thy face . . . eat bread." Yet for all of the troubles that farming at first caused the Natufians and their descendants, they had no choice but to adopt it, says Bar-Yosef. And once they did, there was no turning back: Farming produces more food per acre than hunting and gathering, and the sedentary life led to a population boom that could then be supported only by tilling the earth. As the world began to warm again about 10,000 years ago, the Natufian sites suddenly ballooned to nearly 25 times their previous size.

A yearning to return to an Eden-like existence seems to be an unspoken theme among some environmentalists, who see modern human civilization as a blight on Earth's landscape. But as the new findings on climate and human evolution demonstrate, there was no point in time when humans and Mother Nature lived in perfect harmony. From walking on two legs to making the first stone tools, from harnessing fire to settling down in farming villages, humans have always had to change their ways to cope with changing climate.

While there may be comfort in the fact that the human species has survived large climatic shifts in the past, the ultimate lesson for the conferees gathering in Rio this week is far more complex. Whether the Earth's climate will shift dramatically in the near future as a result of human activities remains uncertain. But if the new research on climate and human evolution reveals anything, it is that if Earth's climate does radically change, the way humans live is likely to change in radical ways, too.

THOUGHT EVOKERS

- Why do some scientists think that a giant meteor was, in part, responsible for the extinction of the dinosaurs?

- How did the skyward thrusting of the Himalaya Mountains affect the world's climate?

III

Principles of Evolution

> If we do discover a complete theory (of everything)... we shall all, philosophers, scientists and just ordinary people, be able to take part in the discussion of why it is that we and the universe exist. If we find the answer to that, it would be the ultimate triumph of human reason ... for then we would truly know the mind of God.
>
> *Stephen Hawking*

In pursuing their examination of older scientific constructs, biologists continue to assess the neo-Darwinian theory of evolution through natural selective processes. Many who accept Darwin's basic premises are attempting to modify some of his ideas about the evolution of organic diversity. However, evolutionary theory remains one of the most viable tenets of recent scientific history.

One part of this tenet—speciation—involves two basic premises: geographical isolation, which prevents reproduction between different groups, and natural selection. Geographical isolation is believed to produce a chance-induced perturbation of the founder genome, and natural selection realigns the genetic variants produced by the perturbation. The growing debate regarding human ancestors reflects the deep controversy over the practice of using fossils as ancestral "stepping stones" to more recent species. As William H. Kimbel of the Institute of Human Origins in Berkeley, California, states, "The really interesting question is not whether H. erectus existed," but that for "the first time in years, we're taking a step back and asking about the theories that underlie our work and the units we use to establish evolutionary relationships."

Evolution and creationism are two countering theories that continue to generate controversy between factions supporting these supposedly opposing premises. Regarding the fundamental question of human origin, religion and science have historically found little ground to agree upon, for the three major Western religions—Christianity, Judaism, and Islam—all teach that the universe was produced by a divine being, and that acceptance of this credo is based on faith. Recently, however, some theologians are attempting to "make room for biological data," and adherents of both concepts have begun to bridge the historic gap between religion and science. Although proponents from both camps will probably never totally agree on all issues, creative dialogue is taking place in the attempt to supply some answers to one of society's most perplexing questions.

> I look at everything as resulting from designed laws, with the details left to the working out of what we may call chance. Not that this notion at all satisfies me. I feel most deeply that the whole subject is too profound for the human intellect.
>
> *Charles Darwin*

8

The Creation

On the question of human origins, religion and science have shared little common ground since the time of Darwin. But now some theologians are trying to make room for biological data.

By Jeffery L. Sheler with Joannie M. Schrof

It is a primordial mystery that has engaged the human imagination from the dawn of time: Who are we? Where did we come from? What is our ultimate destiny? The story of creation—whether told through the imagery of ancient myth, the revelations of sacred tradition, or the theories of modern physics and biology—is the story of what it means to be human. And it is a story over which religion and science have been fighting for centuries. Since Copernicus overturned the church-sanctioned view of Earth as the center of the universe and Charles Darwin posited random mutation and natural selection as the real creators of human life, the biblical view that "In the beginning God created the heavens and the Earth" has found itself under increasing attack in modern Western thought.

Despite the dominance of Darwin's theory—that human beings evolved from lower life forms over millions of years—theologians and religious leaders have yielded relatively little ground on what for them is a fundamental doctrine of faith. The three major Western religions—Christianity, Judaism, and Islam—all teach that the universe is the handiwork of a divine creator who has given humanity a special place in that creation, although the details of just how and when it all occurred are widely disputed. This view of human origins is shared by most Americans

as well. A new poll by the Gallup Organization last month found that nearly half of all Americans subscribe to a fairly literal reading of the biblical creation account, while another large segment believes God played at least some creative role in the universe.

The apparent conflict between religious and scientific explanations of creation has left a centuries-old legacy of suspicion and outright acrimony that in modern times has erupted in open warfare in the nation's courtrooms and classrooms. Often, the modern debate has amounted to little more than a shouting match between extremists on both sides—religious fundamentalists who dismiss evolution as a Satanic deception, and atheistic naturalists who assert that science offers the only window on reality and who seek to discredit religious belief. At least one small denomination, the 400,000-member Christian Reformed Church, is on the verge of a schism over the issue. Conservatives are set to pull out, in large part because of the teaching of evolution in science classes at Calvin College, the denomination's liberal-arts college in Grand Rapids, Mich. "It's just another symptom of the low view of scripture in the denomination," says the Rev. Audred Spriensma of South Holland, Ill., an officer in one of the conservative groups breaking from the church.

Bridge building. Now a move to soften the rhetoric, and perhaps even to bridge the historic gap between

religion and science, is gaining momentum. While few experts suggest an actual convergence of the two is possible, and some question whether it is desirable, creative dialogue is on the upswing. Books and scholarly articles by scientists and theologians exploring the possibilities of a closer relationship have begun appearing with increasing regularity. New organizations are forming and others are expanding whose aim is rapprochement between science and religion. At least 72 organizations worldwide, many in the United States, now provide forums for creative exchange of religious and scientific perspectives. Pope John Paul II, in a 1988 message, heartily endorsed such interaction. Science, said the pontiff, "can purify religion from error and superstition," while religion "can purify science from idolatry and false absolutes."

Although religion and science have a long history of conflict, they have not always been at odds. In the West, early science grew out of a decidedly religious impulse: to understand God and his relationship with man. The biblical picture of an orderly creation by a dependable God gave impetus to scientific inquiry. The universe "made sense" because it was overseen by a Supreme Intelligence who made mathematical description and prediction possible. Throughout Christianity's early history, theologians like Augustine and Thomas

Aquinas did most of the scientific pondering, and their inquiries were seen as a religious quest.

Not until the 15th century did the relationship between science and religion begin to seriously fray. Nicolaus Copernicus, an astronomer who also was the canon of his local cathedral, put forth the idea that Earth did not sit motionless at the center of the cosmos but revolved around the sun. He

---

Scientists & Their Gods:
Stephen Hawking
If we do discover a complete theory (of everything) . . . we shall all, philosophers, scientists and just ordinary people, be able to take part in the discussion of why it is that we and the universe exist. If we find the answer to that, it would be the ultimate triumph of human reason . . . for then we would truly know the mind of God.

---

was so fearful of the religious implications and how the church might respond that he passed his results around anonymously. Only on his deathbed did Copernicus dare sign his name to the astonishing finding. It was Italian astronomer Galileo, building upon Copernicus's work, who argued in the 17th century that scientific inquiry should be free from the restraints of church authority. The church responded by putting him under house arrest and forbidding him to write or speak of Earth's movement around the sun.

Other scientists of Galileo's day struggled privately with conflicts between religious belief and science. But by the 18th century, Enlightenment thinkers were arguing that human reason and scientific empiricism, rather than religious belief, were best equipped to explain human existence. A century later, Charles Darwin, a lifelong member of the Church of England, published "Origin of Species," detailing his theory of evolution. But it was Thomas Huxley, an associate of Darwin's, who

expounded on the theological implications of evolution—that man was not a unique creation of God at all but derived from animal ancestors.

School battles. Church reaction against such thinking was strong from the start, but nowhere so strong as in the United States. Fundamentalists pressed for state laws banning the teaching of evolution in public schools. Literal creationists reached their zenith with their victory in the famous Scopes "monkey trial" in Tennessee in 1925, and evolution almost disappeared from high-school textbooks for a quarter century. In the 1950s, however, Darwin's theory began once again to be incorporated into some biology studies and, by the end of the 1960s, state laws banning evolution texts had disappeared. More recently, court battles have been fought in Arkansas and Louisiana over whether the public schools ought to teach "scientific creationism"—the argument that geologic and fossil evidence is consistent with the Book of Genesis. So far, courts have turned back all such moves.

Today, while some scientists and theologians are sending out peace feelers, literal creationists continue to draw battle lines. For fundamentalists and some evangelical Christians, a literal interpretation of creation in the book of Genesis rules out any chance of rapprochement with science—at least on questions of human origins. "Anyone who believes in evolution cannot accept the biblical record of

---

Scientists & Their Gods:
Paul Davies
My feelings about God and the universe have come about entirely through my science. I hesitate to use the word "God," but in my studies of the universe I have come to the conclusion that there is some purpose to it. The universe has organized itself in such a way as to become aware of itself. As conscious beings, we are part of that purpose.

---

creation," insists Duane Gish, vice president of the Institute for Creation Research in San Diego, a group that espouses scientific creationism.

Under their theory of "recent special creation," taught at hundreds of church-supported schools and Bible colleges throughout the country, God created the universe out of nothing in a series of sudden supernatural acts described in the first two chapters of Genesis. By tracing the biblical genealogies, some adherents estimate that the earth came into being just 10,000 years ago. Fossils and geological evidence of a much older earth, they explain, are the result of Noah's flood or, alternatively, are simply part of an "appearance of age" that God built into the universe. That, presumably, also explains how stars millions of light-years away could be visible on earth if the universe was created so recently. But to many theologians, it also raises the problem of a God engaged in deception.

Many theologians and most scientists—including many who are devoutly religious—dismiss creation science as engaging more in polemics than in rigorous scientific inquiry. Among groups that reject literal creationism is the American Scientific Affiliation, composed of about 2,300 scientists who identify themselves as evangelical Christians. They support a more conventionally scientific approach to the question of origins, acknowledging the evidence for natural processes at work in the universe yet affirming a belief in God as creator and sustainer of life. "A lot of educated people in the sciences see the two as perfectly compatible," says Robert L. Herrmann, a molecular biologist and ASA's executive director.

In fact, many religious scientists are convinced there should be no conflict at all between science and religion. "We have two databases: scripture and nature," says Dorothy Chappell, chairman of the biology department at Wheaton College, a Christian liberal-arts school in Illinois. "Many of us believe God has revealed himself in both." Thus, when science and theology seem to conflict, both should be open to testing and revision. It certainly does not mean scripture is wrong, says Chappell, but it may well mean the interpretation is faulty.

What's often lost in the modern acrimony about teaching evolution and creation science is that theologians themselves have a long history of grappling intellectually with the biblical creation account. Prof. Davis A. Young of Calvin College notes that the fifth-century theologian Augustine warned against taking the six days of Genesis literally. Writing "On the Literal Meaning of Genesis," Augustine claimed that while God created everything in the beginning, some things were made in fully developed form and others were made "in a potential form," says Young, "so that in time they might become the way we see them now." Augustine also worried about the church's public image and took a dim view of interpreting scripture in ways that conflicted with evidence in nature. "The shame," wrote Augustine, "is not so much that an ignorant individual is derided but that people outside the household of faith think our sacred writers held such opinions."

"Geological days." Today, a growing number of theologians, harking back to Augustine, are convinced that more nuanced views of the biblical creation account are required to accommodate scientific evidence. One such attempt is the so-called day-age theory—the idea that creation was gradual, occurring perhaps over millions of years and that the "days" of Genesis actually were geological ages. Proponents of this view cite the Second Epistle of Peter in the New Testament as saying, "With the Lord, one day is as a thousand years." But some Bible scholars object, saying that references in Genesis to "evening and morning" in each of the six days of creation clearly connote 24-hour days.

A similar approach is the "multiple-gap theory," which suggests that creation was a series of instantaneous acts occurring over six 24-hour days but that each was separated by long periods of time. Proponents of this view say it accounts for the sudden

appearance of new life forms, some of which are found in the fossil record. But some text experts see no scriptural basis for long lapses between the six days of Genesis.

The "gap and restitution theory" postulates that the passage of an indeterminate, but presumably very long, period of time is implied between the first and second verses of Genesis:

1. *In the beginning God created the heavens and the earth.*
2. *And the earth was formless and void, and darkness was over the surface of the deep . . .*

This view assumes that there was an initial creation in verse 1, followed by destruction and chaos in verse 2—all of which preceded the "recreation" of an inhabitable earth as spelled out in the succeeding verses. This scenario, its proponents argue, could account for geological evidence for an ancient earth and perhaps even the fossil record. The view finds little support

---

Scientists & Their Gods:
Carl Sagan
The idea that God is an oversized white male with a flowing beard who sits in the sky and tallies the fall of every sparrow is ludicrous. But if by "God" one means the set of physical laws that govern the universe, then clearly there is such a God. This God is emotionally unsatisfying . . . it does not make much sense to pray to the law of gravity.

---

outside of fundamentalist circles.

More prevalent among religious scientists, though less so among conservative theologians and clergy, is "theistic evolution"—the view that evolutionary theory is basically correct and that life on earth, including humanity, evolved over millions of years. But unlike the naturalists, theists consider the evolutionary process, like all other physical processes detectable by science, to be divinely governed. In this view, held by much of mainline

Protestantism, Reform and Conservative Judaism, and Roman Catholicism, the Genesis account is understood as speaking metaphorically of the rela-

---

Scientists & Their Gods:
Henry F. Schaefer
The significance and joy in my science comes in those occasional moments of discovering something new and saying to myself, "So that's how God did it." My goal is to understand a little corner of God's plan.

---

tionship between God and creation, rather than as a scientific or historical account of how and when creation occurred. To that way of thinking, observes Ian Barbour, professor emeritus of science and religion at Carleton College in Minnesota, "we can look at the Big Bang and subsequent evolution as God's way of creating."

A few theologians have begun to argue provocatively that the weight of scientific evidence for evolution is so overwhelming that it calls for a radical reconstruction of the very concept of God. Gordon Kaufman, for example, a professor of theology at Harvard Divinity School, proposes rejecting the theistic view of a Creator as an entity apart from creation and replacing it with the concept of a "serendipitous creativity"—a creative force that reveals itself throughout history and throughout the universe. This theological perspective comes very close to the way many scientists use the word God—to capture the elegance of the natural forces that govern the universe. Other theologians argue that evidence of ongoing change in the universe suggests that God also is in a process of change.

To those theologians and church leaders who are looking for a rapprochement with science, there is much more at stake than an accurate uncovering of primordial history. "Science," says Philip Hefner, head of the Chicago Center for Religion and Science, "is the only cultural institution that Christianity has run away from in 2000 years of history." If it is to maintain its credibility, adds Howard J. Van Till, a professor of physics and astronomy at Calvin College, the church must come to terms with science—"not only for the sake of maintaining peace within the church but for the sake of presenting an effective Christian witness to a scientifically knowledgeable world."

In large measure, the covenantal relationship between God and Israel in the Old Testament and the foundational doctrines of the Christian faith—sin, redemption, and salvation—are inextricably linked to the

---

Scientists & Their Gods:
Charles Darwin
I look at everything as resulting from designed laws, with the details left to the working out of what we may call chance. Not that this notion at all satisfies me. I feel most deeply that the whole subject is too profound for the human intellect.

---

biblical concept of Creator and creation. To Van Till and others who subscribe to theistic evolution, that is the real purpose of the Genesis creation story: to declare the relationship between God and his creation. "Creation and evolution are not contradictory," says Van Till. "Rather, they provide different answers to a different set of questions."

Yet it is for much the same reason—defending traditional biblical faith—

that some religious groups are sharpening their attacks on evolutionary science. Phillip Johnson, a Berkeley law professor, in his book "Darwin on Trial," attacks the "dogmatism" of some scientists who, he says, present evolution "as a religion." While attempting to distance himself from creation literalists, he focuses on what he describes as weak spots in the evidence for evolution. Atheistic naturalism, says Johnson, "is not merely the conclusion that neo-Darwinists draw from their scientific theory" but rather "the metaphysical basis of the theory itself."

The stakes for science in the dialogue may not seem as high. Few on either side expect theology to provide new information or useful techniques that will make science more effective or alter the scientific view of human origins. But, say scholars from both fields, science has much to gain on questions of ethics and human values. Science devoid of values, says Rosemary Radford Reuther, theology professor at Garrett-Evangelical Theological Seminary in Evanston, Ill., gave rise to "the demonic uses of technology" in Auschwitz and Hiroshima. Today, she says, "science cannot afford to ignore questions of value." And religion, says Langdon Gilkey, professor emeritus of theology at the University of Chicago Divinity School, "is one of the main places . . . where serious theoretical, practical and ethical reflection goes on."

More pragmatically, some scholars warn, if science strikes an aloof and even derisive posture toward the values and belief system of society, it can-

not expect to continue to find broad-based political and financial support. Scientists "must learn humility," says Mihaly Csikszentmihalyi, professor of psychology at the University of Chicago, "and be ready to see our conclusions as temporary and open to challenge." After World War II, he says, "the scientific community became overly optimistic that science and technology had the power to liberate human beings from the mental shackles that old-fashioned religion, political ideology and morality had imposed." But, he says, they missed "an important aspect of human psychology that earlier religious approaches had recognized: that left to its own devices . . . human consciousness is typically in a state of chaos and conflict." Though that's not exactly "empirical evidence for original sin," says Csikszentmihalyi, "it is clear that the human psyche is by nature more disordered" than some optimistic scientists would have it.

Just how much science and religion are willing or able to learn from one another is still far from certain. Indeed, centuries of intellectual chauvinism and mutual distrust may ultimately prove to be simply too great an obstacle to productive dialogue, and an important opportunity may be lost. Good science, Albert Einstein once wrote, is created "only by those who are thoroughly imbued with the aspiration towards truth and understanding." The same must surely apply to "good theology."

THOUGHT EVOKERS

- Historically, what have been the major differences in thought between creationism and the scientific concept of evolution?

- Are the concepts of evolution and creationism absolutely opposed to each other? Explain.

9 *How Did Life Start?*

*Earth could have been just another empty chunk of rock.
But something happened here,
and it may have taken place on a stage of clay.*

By Peter Radetsky

On a calm, clear day in February 1977, Jack Corliss and two fellow explorers wedged themselves into the tiny, cramped cabin of the research submarine *Alvin,* said good-bye to the two support ships at the surface, and began a long descent into darkness. About 90 minutes later *Alvin* was gliding along the seafloor a mile and a half below the surface of the Pacific, and Corliss, a burly Oregon State University marine geologist, was peering out the porthole, searching for a phenomenon that had been suspected but never seen: submarine hot springs.

Searchlights blazing, *Alvin* cruised through black water above the Galápagos Rift, an undersea volcanic ridge along the equator 200 miles west of Ecuador. It was in just such a place, Corliss and the others surmised, that these so-called hydrothermal vents would be found—if they existed. Suddenly, just ahead, they spotted a huge cluster of clams. That was odd. What should so many large clams, fully a foot long, be doing so far below their sources of food?

Alvin floated nearer and Corliss pressed closer to the porthole. "I saw a veil of shimmering water," he recalls. "It reminded me of the way air wavers above hot pavement." *Alvin* extended its mechanical arm, in its grip a thermometer; 44 degrees. Not particularly warm by terrestrial standards, but at the ocean floor, where the climate is ordinarily close to freezing, this was bathtub temperature. The crew of the sub broke out in a

cheer. The glistening veil was actually a sheet of water rising from the rocky floor. Corliss and his team had found their submarine hydrothermal vent.

As it turned out, Corliss's team found four more hot springs in the Galápagos Rift, and since that initial exploration numerous other vents have been detected elsewhere on the ocean bottom. In these vents, seawater percolates through a maze of mineral-lined fractures to encounter magma deep in Earth's crust. Heated by the magma, the water rushes upward to escape in the kind of shimmering veil Corliss noticed or in turbulent "black smokers." The team also encountered strange forms of life clustered around the vents—giant worms and blind

---

Questions about the origin of life are as old as Genesis and as young as each new morning.

---

white crabs scurrying over bulging pillows of lava rock. This vast submerged world had a bizarre, elemental quality to it, as though it were a holdover from primordial Earth.

"It was totally amazing," Corliss says. "I began to wonder what all this might mean, and this sort of naive idea came to me. Could hydrothermal vents be the site of the origin of life?"

Questions about life's origin are as old as Genesis and as young as each new morning. For scientists, there are no definitive answers. But if no one has yet pinned down the secret, it hasn't been for lack of trying. Those

investigating the origin of life are a rambunctious, scrappy group, in which no two people see things quite the same way, and it doesn't help that it's awfully tough to prove or disprove any particular contention. After all, how can you really know what happened when Earth formed 4.6 billion years ago? Two things these scientists can agree on, however, are that the first kinds of life, whatever they were, must have been able to reproduce themselves and must have carried information.

Self-replication is the cornerstone of any definition of life. Birds do it, bees do it; certainly our evolutionary forebear, no matter how simple an organism, must have been able to do it. To sustain life, information about oneself must be passed from one generation to the next. It is that information, in the form of inheritable characteristics, that gives life continuity. It is the accidental altering of those characteristics over time that makes evolution possible. We do this with genes. But what is not at all clear is how our ancient ancestors did it, or what form those ancestors took. Evolutionary biologists have traced our family tree to bacteria, one-celled organisms that have been found in rock formations 3.5 billion years old. But even these "primitive" creatures were already quite sophisticated. They had genes of DNA and RNA and were made of protein, lipids, and other ingredients. Something simpler must have preceded them.

A hint as to what that may have been came in 1981, when Thomas Cech of the University of Colorado discovered a kind of RNA that functioned as an enzyme, partially triggering its own replication. Until then, replication had been thought possible only through a collaboration among DNA, the storehouse of genetic information, RNA, the mobile dispenser of that information, and protein, which exclusively makes up the enzymes that catalyze the process. Now Cech had shown that RNA could be an enzyme and therefore could once have taken care of the whole business by itself. The news galvanized scientists, who enthusiastically painted a picture of an ancient world inhabited by naked RNA genes, which went on their way merrily self-replicating until DNA and protein evolved to assist in the procedure. Thus ensued the development of living cells and the very bacteria we claim as our own ancestors.

But while this proposed RNA world was certainly closer to the origin of life, it clearly wasn't the beginning. Although much simpler than bacteria, RNA is still a complicated piece of molecular machinery, containing more than 30 atoms connected in an intricate, interlocking fashion. It couldn't have sprung wholly formed into the primordial landscape. Something preceded it. That something must have been the simple carbon-based molecules that underlie all life—organic compounds.

What were those first organic compounds? And how did they form? The questions bedevil origin-of-life researchers. Over the years they have come up with a host of imaginative and intensely debated possibilities. Perhaps the most influential first surfaced four decades ago, when in a dramatic experiment a University of Chicago graduate student named Stanley Miller simulated the creation of life in a laboratory.

Today Miller is a renowned and feisty 63-year-old professor of chemistry at the University of

California at San Diego. Back in Chicago in 1953, however, he little knew what he was getting himself into. "My research director, Harold Urey, gave a talk about the origin of Earth and the solar system," he recalls. "He said that if you have an atmosphere like that of early Earth you ought to be able to make organic

---

As the scientists watched, fluids "rained" out of the gas chamber, turning the clear "ocean" pink.

---

compounds easily. I said, 'I want to do it,' but he tried to talk me out of it. It was a very risky experiment, and it was his responsibility to make sure that I had an acceptable thesis within a couple of years. I said that I'd give it a try for six months to a year, and if that didn't work out, I'd do something conventional."

Urey agreed, and the two set to work. They designed a glass apparatus consisting essentially of two flasks connected within a closed circle of glass tubing. Miller pumped into the larger flask the gases thought to be present in the early atmosphere: hydrogen, methane, ammonia, and water vapor. The smaller flask he partially filled with water—it represented the primitive ocean. He then shot bolts of electric current through the gaseous mixture to simulate primordial lightning storms. For a week the electricity sparked, while Miller sat back to see what would happen.

"It didn't take long to see I had it," he says. "The organics just poured out. It was very exciting."

As the scientists watched, fluids "rained" out of the gas chamber, turning the clear water in the "ocean" pink, then deep red, then yellow-brown. When Miller analyzed the brew, he found that it contained amino acids, the building blocks of protein. The "lightning" had reorganized the molecules in the "atmosphere" to produce organic compounds. It looked as though making organics was easier than anyone had

suspected. Perhaps the origin of life was simplicity itself, nothing more than the routine consequence of basic conditions on early Earth.

People were stunned. Articles appeared in major newspapers across the country, prompting predictions that, like Dr. Frankenstein, researchers would soon concoct living organisms in their labs. A Gallup poll asked people whether they thought it possible "to create life in a test tube." (Seventy-eight percent answered—perhaps hopefully—no.) And the simple experiment ("It's so easy to do—high school students now use it to win their science fairs," Miller says) stimulated a rush of studies, with the result that a number of other organic compounds, including adenine and guanine, two of the ingredients of RNA and DNA, were produced by similar procedures.

Thus emerged the picture that has dominated origin-of-life scenarios. Some 4 billion years ago, lightning (or another energy source, like ultraviolet light or heat) stimulated a hydrogen-rich atmosphere to produce organic compounds, which then rained down into the primitive ocean or other suitable bodies of water such as lakes, rivers, or even a "warm little pond," as Charles Darwin once suggested. Once there, these simple compounds, or monomers, combined with one another to produce more complicated organics, or polymers, which gradually grew even more complex until they coalesced into the beginnings of self-replicating RNA. With that came the RNA world and ultimately the evolution into cells and the early bacterial ancestors of life.

The picture is powerful and appealing, but not all origin-of-life researchers are convinced. Even Miller throws up his hands at certain aspects of it. "The first step, making the monomers, that's easy. We understand it pretty well. But then you have to make the first self-replicating polymers. That's *very* easy," he says, the sarcasm fairly dripping. "Just like it's easy to make money in the stock market—all you have to do is buy low and sell high." He laughs. "Nobody knows how it's done."

Some would say the statement applies as well to the first "easy" step, the creation of simple organic compounds. For example, what if the primordial atmosphere wasn't anything like the one Miller and Urey imagined? Would it be so easy to produce organics then?

"The Miller-Urey experiment was a strong foundation because it was consistent with theories at the time," says geochemist Everett Shock of Washington University in St. Louis. "The problem is that subsequent research has swept away a lot of those ideas. The Miller-Urey atmosphere contained a lot of hydrogen. But now the atmosphere of early Earth is thought to have been more oxidized."

That makes Miller's scenerio less probable, because it's a lot harder to make organic molecules in the presence of oxygen. A hydrogen-rich atmosphere is relatively unstable. When zapped by lightning or other sources of energy, molecules in that environment readily tumble together into organic compounds. Not so in a heavily oxidized atmosphere. While an infusion of energy may cause a few simple organics to form, for the most part the results are inorganic gases like carbon monoxide and nitrogen oxide. "These are the constituents of smog," says Shock. "So basically what you're getting is a lot of air pollution."

"That's worried people for the last 10 to 15 years," says Christopher Chyba, a planetary scientist based at NASA's Ames Research Center, south of San Francisco. "There seems to be a contradiction between the fact that we're here and evidence that early Earth was not very hospitable to the formation of organics. How do you resolve the dilemma? One way is to take advantage of the fact that asteroids and especially comets are rich in organic compounds. Maybe there was a way that those organics reached early Earth intact."

In other words, maybe the beginnings of life came from interstellar space. The notion is not as far-fetched as it may seem. "If you go to the moon," says Chyba, "or look at the craters on Mars or Mercury, what you see is that the whole inner solar system was being subjected to a very intense bombardment from space at that time. You can infer that the same was true for Earth." And in fact, in the early nineteenth century, organic molecules were found in a meteorite, although some people suspected that it had simply acquired earthly organics in the thousands of years since it had landed. In 1969, however, such skepticism was dispelled once and for all when a meteorite fell in Murchison, Australia. A prompt examination revealed a large number of amino acids, components of RNA and DNA, and other organic compounds.

"More recently," says Chyba, "in 1986, European and Soviet spacecraft flew by Halley's comet. People had strongly suspected that comets were rich in organics, and that was absolutely borne out by the observations made by the spacecraft." And whereas the fraction of organics in meteorites is no more than one-twentieth of their mass, the flybys found Halley to be fully *one-third* organic compounds.

However, says Chyba, it's likely that most organics aboard meteorites and comets never made it to Earth. "At these velocities, at least 10 to 15 miles per second, the temperatures you reach on impact are so high that you end up frying just about everything." And those organics that survived would probably have been too few and too scattered to evolve into life.

But interplanetary dust particles (IDPs for short) are another matter. In contrast to their larger cousins, these particles, tiny specks no larger than .004 inch across, routinely reach Earth. "They get slowed way up in the atmosphere," says Chyba. "Then they

remain floating around for months, even years, before they come down. NASA samples IDPs directly in the atmosphere with modified U2 spy planes fitted with adhesive collectors on the wings." What researchers have found is that IDPs also contain organic material—although only about 10 percent worth. Perhaps, then, dust seeded early Earth with the stuff of life.

Not surprisingly, not everyone thinks so. "If you have to depend on such low amounts of organic material as that found in IDPs," says Miller, "then from the standpoint of making life on Earth you're bankrupt. You're in Chapter Eleven. Because you just don't have enough." His point rests on simple common sense: the greater the amount of organics, the greater the possibility that they would have interacted with one another. Too few organics, and odds are that they could never have gotten together to begin the process of life in the first place. "Organics from outer space," Miller scoffs. "That's garbage, it really is."

There's another possible drawback to the notion of an extraterrestrial origin of life, acknowledged by Chyba himself. "The surface of early Earth would have been a very hostile place," he says. "The biggest impacts would have generated enough heat to evaporate the entire ocean, probably several times. And leaving the biggest impacts aside, the upper tens of meters of the oceans would routinely have been evaporated and the surface of Earth sterilized by these giant impacts."

Where, then, in such a nightmarish environment, could emerging life have been sufficiently protected? The only safe place—safe, at least, after the last total evaporations were over and done with—would have been in the deep ocean. And that, says Jack Corliss, is where hydrothermal vents come into the picture.

Since his discovery of the Galápagos hot springs, Corliss, who now works at NASA's Goddard Space Flight Center, in Greenbelt, Maryland, and a growing number of his colleagues have been promoting the notion that hydrothermal vents were the birthplace of life. "The thing about the hot springs," Corliss says, "is that they provide a nice, safe, continuous process by which you can go from very simple molecules all the way to living cells and primitive bacteria."

The crux is the word *continuous*. For besides providing safe harbor for the development of life, vents offer a natural temperature gradient. The vents have it all, from the "cracking front" in the interior, where temperatures reach 1300 degrees and cool water filtering down from above cracks the superheated rock, to the 40-degree seafloor. "Whatever temperature you want," says Corliss, "you have your choice." And any chemist will tell you that where you find a temperature gradient is where you'll find chemical reactions—maybe even the ones that began life.

The reactions Corliss envisions began at the cracking front, half a mile deep in the planet's crust, where seawater encountered hot magma. There, in this seething caldron, elements like carbon, oxygen, hydrogen, nitrogen, and sulfur interacted to form new, organic compounds. "Just as in the Miller-Urey experiments," says Corliss, "if you heat simple molecules to high temperature, you can make organic compounds."

But heat is a double-edged sword. It facilitates chemical reactions, but it can also destroy the products of those reactions. If exposed to high heat for too long, organic compounds decompose. "It's a very simple argument: if you keep a roast too long in an oven that's too hot, it's going to get charred," says Miller, who has little use for this scenario either. "The vent hypothesis is a real loser. I don't understand why we even have to discuss it," he says, his voice rising to an exasperated falsetto.

Corliss, however, thinks he has an ace in the hole: a vent's temperature gradient. He thinks it likely that the circulating seawater cooled the newly formed compounds almost immediately. "If you quenched them very rapidly, you could preserve them," he says. "Then they rose and mixed and worked their way up in the hot springs, through this huge complex of fractures, cooling as they went."

Finally the organic compounds were deposited onto the clay minerals lining the mouth of a vent. And there they stayed. Rather than simply emerging and dissipating into the vast ocean where they might never encounter another organic molecule, the compounds accumulated on the clay surface. There, in a concentrated colony, they were able to interact with one another and with the endless supply of new compounds rising in the hot springs, until over time the first stirrings of primitive life emerged.

"It's the perfect environment," Corliss says. "You couldn't design it better. With the clay minerals lining the fractures in the upper part of the hot springs, the organic material has something to stick to. It's an ideal way to concentrate the organic material made at the cracking front. Now it can build up and evolve."

The prospect is bolstered by the likelihood that in the turbulent early Earth there were many more hydrothermal vents than today. "Presumably it was hotter within primitive Earth, so there was even more hydrothermal circulation to cool things down," says Everett Shock. And, therefore, more safe havens in which life might have evolved.

Furthermore, the clay lining the vents could have been far more than just a convenient medium on which organic compounds could evolve. Chemist A. Graham Cairns-Smith of the University of Glasgow sees clay as a solution to the mystery of how simple organics made the leap all the way to self-replicating genetic material. In fact, Cairns-Smith sees clay itself as the first genetic substance, what he calls a crystal gene.

Clay minerals, he explains, are crystals formed from the weathering of rocks by water. And clay, like any crystal, grows by itself—think of crystals of frost expanding on a windowpane. Crystals, in other words, self-replicate. So if self-replication is the key, life did not start with organic molecules. Life started with crystals. That is, it started with clay.

It's not a new idea—the bible proposed it long ago, in a slightly different form. But in Cairns-Smith's hands the notion takes on an evocative modern flavor. "With clay, I'm advocating an earlier genetic material that is fundamentally different from DNA and RNA," he says. "You needed a previous stage of evolution in which the present means of evolution was itself evolving."

Again, picture a hydrothermal vent, with organic compounds settling on clay crystals lining the fissures. But this clay was no inert surface upon which organic reactions happened to take place—it was living, growing, even assisting those reactions. As the crystals grew, they developed nooks and crannies that were a perfect fit for the organic molecules rising in the swell of water. As snugly as pegs settling into holes in a pegboard, these molecules made themselves at home in this surface. Once there, they reacted with other molecules comfortably ensconced in niches next door. Because the positioning was so precise, similar reactions could occur over and over again. The crystals, then, actually catalyzed the formation of new organic compounds.

In time the organics evolved into RNA, which, with its strong interlocking structure, returned the favor, helping out the growing clay crystals. "I don't think RNA's genetic function came first," Cairns-Smith says. "My guess is that at first it had a structural function. It helped stick the crystals together." Finally, as it became a self-replicating molecule, RNA jettisoned its clay scaffolding. And early life struck out on its own.

This scenario, attractive as it may seem, is—like so many others—too farfetched for Miller. "It's not that I don't want to entertain new ideas—that's fine," he says. "The question is, does this chemistry work? Actually work in the lab? Either it does or it doesn't." His point is well taken. Whatever else may be said about Miller's ideas, his experiments worked. Talk, even informed talk, is cheap. If they're to have an impact comparable to Miller's, these champions of crystals and vents and interstellar particles must *demonstrate* their scenarios.

But how? You can't try to make early life at existing hot springs—they're already replete with bacteria and other life-forms, so the environment just can't be the same as it was on the primordial planet. And re-creating an ancient hydrothermal vent in the lab is a mind-boggling prospect. Still, vent researchers are busily conducting experiments designed to do just that. Elsewhere, Chyba is collaborating with Carl Sagan and others in an attempt to nail down the possible link between extraterrestrial objects and the origin of life. And Cairns-Smith is investigating the chemical relationships between minerals and organic compounds.

But while he recognizes the importance of experimental proof, Cairns-Smith cheerfully acknowledges that he may never come up with any. "I'm hoping that people with new techniques or people who make the appropriate discoveries will phone me up and say, 'By the way . . .' The origin of life depended on all sorts of accidental circumstances. Proving how it happened will take another piece of luck."

THOUGHT EVOKERS

- Describe the experiment that Stanley Miller used to simulate the creation of life.

- Discuss the current hypothesis regarding the "clay" involvement in the creation of life.

- What is the cornerstone of any definition of life? Explain.

Erectus Unhinged

*Debate over a human ancestor reflects deeper splits
concerning the nature of fossil species*

By Bruce Bower

For more than 40 years, anthropologists have generally agreed that *Homo erectus* served as an evolutionary link between our earliest direct ancestor, *Homo habilis*, and modern *Homo sapiens*. This view holds that a hardy breed of *H. erectus* spread from Africa to Asia and Europe and lived from approximately 1.8 million to 400,000 years ago.

But in the last few years, *H. erectus* has suffered an identity crisis. Leading investigators now propose three contrasting theories of human evolution that would give any ancient ancestor cause for concern. One proposal advocates sticking with a single, widespread *H. erectus*; another calls for splitting *H. erectus* into at least two species, only one of which evolved into modern humans; and a third seeks to abolish *H. erectus* altogether, placing its fossil remains within an anatomically diverse group of *H. sapiens* that split off from *H. habilis* about 2 million years ago.

Disagreements of this sort stem from a fundamental parting of the ways about how to discern a species in the fossil record. Most anthropologists accept the species as the basic unit of evolution, while acknowledging that defining a species, even among living animals, often presents problems. Thus, different theories about how best to sort out extinct species based on the features preserved in ancient bones fuel the dispute over *H. erectus* and other members of the human evolutionary family, known as hominids.

However, some researchers stand outside the fray, viewing any attempt to nail down fossil species as an unscientific, arbitrary exercise in cataloguing the ambiguous bits of anatomy surviving in fossil bones.

"There's a growing diversity as to how species are perceived in modern

> **But in the last few years, *H. erectus* has suffered an identity crisis. Leading investigators now propose three contrasting theories of human evolution that would give any ancient ancestor cause for concern.**

and ancient populations," asserts Erik Trinkaus of the University of New Mexico in Albuquerque. "[Researchers] often end up talking past each other."

In April, Trinkaus and others debated various approaches to understanding *H. erectus* and fellow hominid species at the annual meeting of the American Association of Physical Anthropologists in Las Vegas and in interviews with SCIENCE NEWS.

The roots of this sometimes confusing clash extend back 100 years, when the first *H. erectus* fragments turned up in Java. Initially classified as Pithecanthropus, or ape-man, these Asian specimens and most ensuing hominid finds received a unique species designation from their discoverers. In the early 1950s, anthropologists realized that human evolution made no sense if virtually all fossil

discoveries represented different species. Taking the view that an ancestral species with a wide array of skeletal features gradually transforms into a descendant species, researchers proceeded to group fossils into a much smaller number of species.

So-called "lumping" of specimens led to a picture of human evolution as a series of three progressive steps, with *H. habilis* begetting *H. erectus* begetting *H. sapiens*.

But by the early 1980s, these ancestral lumps had begun to stick in the throats of some anthropologists. At the same time, concern grew that the definition of a species used by biologists and often borrowed by anthropologists—namely, characterizing a species as a group of organisms that reproduce only among themselves—offered no help in evaluating fossils.

Another approach—called cladistic, or phylogenetic, analysis—rapidly gained popularity. This view holds that new species evolve relatively quickly rather than in a series of gradual adjustments within ancestral species. Specifically, cladistics assumes that although most members of a population of related organisms display the "primitive" skeletal features that arose early in their evolutionary history, some members of the population sport "derived," or advanced, anatomical features that appeared later. A consistent pattern of unique derived features on a group of fossils serves as a species marker.

Phylogenetic studies indicate that *H. erectus* fossils actually encompass two species, one in Asia that became extinct and another in Africa that evolved into modern humans. Peter Andrews of the Natural History Museum in London reported in 1984 that most skeletal features commonly accepted as unique derived traits of *H. erectus* are actually primitive retentions shared by earlier *Homo* species. Moreover, the seven derived characteristics exclusive to *H. erectus* appear predominantly among Asian fossils. These include an angling of the cranium that produced a bony ridge at the top of the head, thick cranial bones, a cleft in the bone just behind the ear, and a plateau-like bony swelling at the back of the head.

Since these features appear in only one geographically restricted set of fossils and do not turn up later in modern humans, Andrews suggests that Asian *H. erectus* met extinction on a side branch of human evolution. A separate species of African hominids living at the same time evolved into *H. sapiens,* he posits. Andrews' analysis dovetails with the theory that modern humans originated in Africa around 200,000 years ago and then spread throughout the world.

Bernard Wood of the University of Liverpool, England, has elaborated on Andrews' phylogenetic thesis. In the Feb. 27 NATURE, Wood presents a cladogram—a tree diagram organizing hominid species according to the number of derived features shared by groups of fossils—based on analysis of 90 cranial, jaw, and tooth measurements. Wood concludes that, sometime before 2 million years ago, at least three *Homo* species emerged in Africa: the relatively small-brained *H. habilis*; a group with larger brains and teeth, which he calls *H. rudolfensis*; and *H. ergaster,* represented by the fossils that Andrews separated from Asian *H. erectus.*

The three species apparently shared an unidentified common ancestor, with *H. ergaster* serving as the precursor of *H. sapiens,* Wood argues.

Wood splits up early *Homo* species in a reasonable way, notes Ian Tattersall of the American Museum of Natural History in New York City. But neither phylogenetic theory nor any other approach offers practical help to fossil species hunters, Tattersall maintains. Closely related living primate species often differ in only one or a few subtle anatomical features, which may not show up in a set of bones, he points out (SN: 4/13/91, p.230). Thus, cladistic analysis tends to lump together some hominid species that share derived anatomical characteristics, he holds.

In an article accepted for publication in the JOURNAL OF HUMAN EVOLUTION later this year, Tattersall advises investigators to use the phylogenetic approach to identify groups of fossils with derived features that signal either a distinct species or possibly a clutch of related species. Lumping inevitably occurs, but the general pattern of human evolution remains unobscured, he argues.

H. sapiens also requires splitting when viewed under this modified phylogenetic lens, Tattersall contends. He places several partial skulls found at European sites and usually assigned to early, or "archaic," *H. sapiens* (mostly dating to around 200,000 to 400,000 years ago) in a new species, *H. heidelbergensis.*

"It's a virtual certainty that speciations have been much more common in hominid biological history than many paleoanthropologists have been willing to admit," he asserts.

Tattersall has it exactly backwards, according to adherents of the theory of "multiregional evolution." The phylogenetic approach fails to appreciate the anatomical diversity that arises within different populations belonging to the same species, argues Milford H. Wolpoff of the University of Michigan in Ann Arbor. Wolpoff and his colleagues champion an evolutionary perspective in which each hominid species encompasses one or more populations that share the same common ancestor, follow the same evolutionary patterns over time, and yield anatomical evidence of a historical beginning and end.

H. erectus clearly splits off from *H. habilis,* but it gives no sign of an evolutionary demise, according to a study conducted by Wolpoff and Alan G. Thorne of Australian National University in Canberra. Instead, *H. erectus* gradually merges into the range of skeletal characteristics observed in regional populations of early *H. sapiens,* Wolpoff and Thorne argue. Of the 23 derived anatomical

---

In other words, *H. erectus* never existed and *H. sapiens* has evolved in several parts of the world for approximately 2 million years, Wolpoff and Thorne maintain.

---

traits that distinguish *H. erectus* from *H. habilis,* 17 consistently turn up on *H. sapiens* fossils, they assert.

In other words, *H. erectus* never existed and *H. sapiens* has evolved in several parts of the world for approximately 2 million years, Wolpoff and Thorne maintain.

Evolutionary patterns observed in four different regions—Africa, Europe, China, and Australia-Indonesia—show continuous, gradual change from about 2 million years ago to the most recent human populations, with no evidence of Africans replacing the other groups, Wolpoff and Thorne contend. They also hold that *H. sapiens* encompasses most, perhaps all, specimens now classified as Neandertal (SN: 6/8/91, p. 360).

The few anatomical idiosyncrasies separating *H. sapiens* from fossil remains widely attributed to *H. erectus*—such as greater cranial volume, smaller teeth, and lighter limb bones—reflect evolutionary trends in the former species toward larger brains and a greater reliance on tools and other technologies spawned by increasing cultural complexity, Wolpoff argues.

In Wolpoff's view, the merging of *H. erectus* into *H. sapiens* (first proposed in the 1940s by German anatomist Franz Weidenreich, who continues to inspire the multiregional approach) forces scientists to take a closer look at anatomical changes that have occurred over time within our species. It also exposes the need for a workable definition of "anatomically modern humans," he says.

Between those vying to split or to sink *H. erectus* stand some stalwart defenders of its status as a unified species. "I see *Homo erectus* as a single species that spread across the Old World," says G. Philip Rightmire of the State University of New York at Binghamton. *H. erectus* probably gave rise to modern humans in a restricted geographic area, for example Europe, where temperatures cooled dramatically around 400,000 years ago, or possibly in Africa, Rightmire suggests. *H. erectus* populations apparently survived for a while in Asia, whereas *H. sapiens* thrived elsewhere, he says.

To buttress his theory, Rightmire offers a reassessment of a group of fossil skulls and skull fragments found at the Ngandong site in central Java. Multiregional theorists such as Wolpoff view the anatomy of these skulls as intermediate between *H. erectus* and *H. sapiens*, indicating a long,

gradual evolution toward modern humans in that part of the world.

However, the Ngandong fossils—poorly dated, but generally placed between 100,000 and 250,000 years old—clearly fall within the range of anatomy observed in older *H. erectus* skulls from Java and elsewhere, Rightmire contends. This holds for the size and shape of Ngandong braincases, the thickness of the cranial bones, and other features, he points out.

In contrast, the earliest *H. sapiens* specimens display marked increases in brain size, changes in cranial bones that signify shifts in brain organization, and a more flexed cranial base, indicating a vocal tract capable of producing a greater variety of speech sounds—all signs of substantial genetic changes that produced a new species in a relatively short time, Rightmire holds.

Another study, conducted by Steven R. Leigh of Northwestern University in Evanston, Ill., lends some support to Rightmire's contention that a measurable split occurs between *H. erectus* and *H. sapiens*. Leigh examined 20 *H. erectus* skulls from Africa, China, and Indonesia that span a broad time range, as well as 10 early *H. sapiens* skulls. Significant expansion of brain size from the oldest to the most recent specimens occurs in the latter group, whereas the three regional samples of *H. erectus* show no such increases, Leigh reports in the January AMERICAN JOURNAL OF PHYSICAL ANTHROPOLOGY.

However, analysis of the Chinese and Indonesian skulls reveals substantial brain-size increases that do not necessarily coincide with Rightmire's view of an anatomically stable *H. erectus* inhabiting the entire Old World, Leigh points out.

The single-species view gets further ammunition from another study of 70 hominid craniums, mainly *H. erectus* and *H. sapiens* specimens. The seven derived features considered unique to

Asian *H. erectus* by Peter Andrews also appear on many African fossils attributed to *H. erectus*, as well as on a significant number of *H. habilis* and early *H. sapiens* specimens, according to Gunter Brauer of the University of Hamburg, Germany, and Emma Mbua of the National Museums of Kenya in Nairobi.

Although additional anatomical features need study, cladistic procedures mistakenly assume that unique derived traits are either present or absent in all members of a species, Brauer and Mbua contend in the February JOURNAL OF HUMAN EVOLUTION. They emphasize Tattersall's point that the same derived features may occur to a greater or lesser extent in different hominid species. Investigators need better data on variations in the skeletal anatomy of living primates and fossil hominids, they conclude.

Some anthropologists take a dim view of the entire controversy surrounding hominid species. "These fights over species classification are somewhat of a waste of time," says Alan Mann of the University of Pennsylvania in Philadelphia. "Most researchers see *Homo erectus* as a single species that evolved into *Homo sapiens*."

Others argue that fossil bones provide too little evidence for teasing out hominid species.

"Fossil species are mental constructs," contends Glenn C. Conroy of Washington University in St. Louis, who directed an expedition that recently found an approximately 13-million-year-old primate jaw in southern Africa (SN: 6/29/91, p.405). "Cladistic approaches try to separate species out of a vast array of biological variability over a vast time range, and I don't think they're capable of doing that."

Conroy prefers to group hominid fossils into "grades," or related groups tied together by general signs of anatomical unity with no evidence of sharp breaks between species. Thus, an *Australopithecus* grade (which includes the more than 3-million-year-old "Lucy" and her kin) merges into a grade composed of *H. erectus* fossils and then shades into a *H. sapiens* grade, in Conroy's view.

"I'd put our limited funding into looking for new fossil primates or studying living primates, rather than pushing cladograms or arguing about the number of *Homo* species," he asserts.

But anthropologists wrangling over *H. erectus* and other hominid species find room for optimism amid their discord.

"The really interesting question isn't whether *H. erectus* existed," remarks William H. Kimbel of the Institute of Human Origins in Berkeley, Calif., a proponent of phylogenetic analysis. "For the first time in years, we're taking a step back and asking about the theories that underlie our work and the units we use to establish evolutionary relationships. It's a healthy sign that we're debating these questions vigorously."

THOUGHT EVOKERS

- Explain three contrasting theories regarding human evolution.

- Describe what is meant by the phrase *cladistic analysis*.

- Why does G. Philip Rightmire consider *Homo erectus* a single species?

IV Evolution and Diversity

For years ecologists have been warning us of impending environmental crisis. The rapid increase in the human population, especially when coupled with advanced technology, has resulted in increased degradation of the biosphere. For centuries humans have ignored the fact that we, like all life forms, are only part of an interlocking system of biotic and abiotic components. This ignorance, fueled in large part by greed, has led to an unimagined decrease in the earth's ecological diversity. Biologists accept the fact that the viability of any ecosystem is based in diversity—reduce ecological diversity, reduce viability. And yet we continue to destroy the diversity of rain forests, oceans, fresh waters, soils, and our atmosphere. We either ignore or will not admit what is happening. We talk of "multiple land use" as a positive economic concept, when in fact it is a euphemism for more environmental simplification and destruction. A case in point: the planet's oceans. Historically one of the earth's most pristine, complex, and important ecosystems, our oceans are losing diversity at an alarming rate. Caused in large measure by the human-induced perturbations of overpopulation, large-scale agricultural practices, decimation of forest ecosystems, and continuing introduction of toxic wastes, the seas have lost much of their ecological viability. Endemic species of the Atlantic, Pacific, and other oceans are being over-harvested. As their numbers and variety decline further, decimation continues because of technological "advances" in harvesting practices. Mile-long drift nets, often lost at sea, are indiscriminate killers, destroying any species that becomes entangled in the mesh: fish, birds, mammals.

From an evolutionary perspective the "earth could have been just another chunk of rock," but a marvelous transformation has occurred, and one prevailing hypothesis is that life may well have evolved on a "stage of clay." According to this idea, life apparently arose through "a long succession of chemical steps that were bound to take place under conditions that also changed through the past four billion years." Without a sense of "human stewardship" regarding the biosphere, these continuing chemical steps may be modified in the future to the point where the decrease in viability and diversity of the earth's species affects the health of the biosphere itself—including the owl, the salmon, and the human.

11 The Beginnings of Life on Earth

Life arose naturally through a long succession of chemical steps that were bound to take place under the conditions that prevailed on earth four billion years ago

By Christian de Duve

Advanced forms of life existed on earth at least 3.55 billion years ago. In rocks of that age, fossilized imprints have been found of bacteria that look uncannily like cyanobacteria, the most highly evolved photosynthetic organisms present in the world today. Carbon deposits enriched in the lighter carbon-12 isotope over the heavier carbon-13 isotope—a sign of biological carbon assimilation—attest to an even older age. On the other hand, it is believed that our young planet, still in the throes of volcanic eruptions and battered by falling comets and asteroids, remained inhospitable to life for about half a billion years after its birth, together with the rest of the solar system, some 4.55 billion years ago. This leaves a window of perhaps 200–300 million years for the appearance of life on earth.

This duration was once considered too short for the emergence of something as complex as a living cell. Hence suggestions were made that germs of life may have come to earth from outer space with cometary dust or even, as proposed by Francis Crick of DNA double-helix fame, on a spaceship sent out by some distant civilization. No evidence in support of these proposals has yet been obtained. Meanwhile the reason for making them has largely disappeared. It is now generally agreed that if life arose spontaneously by natural processes—a necessary assumption if we wish to remain within the realm of science—it must have arisen fairly quickly, more in a matter of millennia or centuries, perhaps even less, than in millions of years. Even if life came from elsewhere, we would still have to account for its first development. Thus we might as well assume that life started on earth.

How this momentous event happened is still highly conjectural, though no longer purely speculative. The clues come from the earth, from outer space, from laboratory experiments, and, especially, from life itself. The history of life on earth is written in the cells and molecules of existing organisms. Thanks to the advances of cell biology, biochemistry, and molecular biology, scientists are becoming increasingly adept at reading the text.

An important rule in this exercise is to reconstruct the earliest events in life's history without assuming they proceeded with the benefit of foresight. Every step must be accounted for in terms of antecedent and concomitant events. Each must stand on its own and cannot be viewed as a preparation for things to come. Any hint of teleology must be avoided.

Building Blocks

The early chemists invented the term "organic" chemistry to designate the part of chemistry that deals with compounds made by living organisms. The synthesis of urea by Friedrich Wöhler in 1828 is usually hailed as the first proof that a special "vital force" is not needed for organic syntheses. Lingering traces of a vitalistic mystique nevertheless long remained associated with organic chemistry, seen as a special kind of life-dependent chemistry that only human ingenuity could equate. The final demystification of organic chemistry has been achieved by the exploration of outer space.

Spectroscopic analysis of incoming radiation has revealed that the cosmic spaces are permeated by an extremely tenuous cloud of microscopic particles, called interstellar dust, containing a variety of combinations of carbon, hydrogen, oxygen, nitrogen, and, sometimes, sulfur or silicon. These are mostly highly reactive free radicals and small molecules that would hardly remain intact under conditions on earth, but would interact to form more stable, typical organic compounds, many of them similar to substances found in living organisms. That such processes indeed take place is demonstrated by the presence of amino acids and other biologically significant compounds on celestial bodies—for example, the meteorite that fell in 1969 in Murchison, Australia, Comet Halley (which could be analyzed during its recent passage by means of instruments carried on a spacecraft), and Saturn's satellite Titan, the seas of which are believed to be made of hydrocarbons.

Christian de Duve, who shared the 1974 Nobel Prize in medicine with Albert Claude and George Palade "for their discoveries concerning the structural and functional organization of the cell," divides his time between the University of Louvain, in Belgium, where he is professor emeritus of biochemistry and founder-administrator of the International Institute of Cellular and Molecular Pathology, and the Rockefeller University, in New York, where he is Andrew W. Mellon professor emeritus. In his latest book, Vital Dust: Life as a Cosmic Imperative, *published by Basic Books in New York, de Duve surveys the entire history of life on earth, from the first biomolecules to the human mind and beyond. The resulting view is of a meaningful universe, in which life and mind are cosmic imperatives. Address: The Rockefeller University, 1230 York Avenue, New York, NY 10021.*

Reprinted by permission of *American Scientist*, Journal of Sigma Xi, The Scientific Research Society.

It is widely agreed that these compounds are not products of life, but form spontaneously by banal chemical reactions. Organic chemistry is nothing but carbon chemistry. It just happens to be enormously richer than the chemistry of other elements—and thus able to support life—because of the unique associative properties of the carbon atom. In all likelihood the first building blocks of life arose as do all natural chemical compounds—spontaneously, according to the rules of thermodynamics.

The first hints that this might be so came from the laboratory, before evidence for it was found in space, through the historic experiments of Stanley Miller, now recalled in science textbooks. In the early 1950s, Miller was a graduate student in the

> **In all likelihood the first building blocks of life arose as do all natural chemical compounds—spontaneously, according to the rules of thermodynamics.**

University of Chicago laboratory of Harold Urey, the discoverer of heavy hydrogen and an authority on planet formation. He undertook experiments designed to find out how lightning—reproduced by repeated electric discharges—might have affected the primitive earth atmosphere, which Urey believed to be a mixture of hydrogen, methane, ammonia, and water vapor. The result exceeded Miller's wildest hopes and propelled him instantly into the firmament of celebrities. In just a few days, more than 15 percent of the methane carbon subjected to electrical discharges in the laboratory had been converted to a variety of amino acids, the building blocks of proteins, and other potential biological constituents. Although the primitive atmosphere is no longer believed to be as rich in hydrogen as once thought by Urey, the discovery that the Murchison meteorite contains the same amino acids obtained by Miller, and even in the same relative

proportions, suggests strongly that his results are relevant.

Miller's discovery has sparked the birth of a new chemical discipline, abiotic chemistry, which aims to reproduce in the laboratory the chemical events that initiated the emergence of life on earth some four billion years ago. Besides amino acids and other organic acids, experiments in abiotic chemistry have yielded sugars, as well as purine and pyrimidine bases, some of which are components of the nucleic acids DNA and RNA, and other biologically significant substances, although often under more contrived conditions and in lower yields than one would expect for a prebiotic process. How far in the direction of biochemical complexity the rough processes studied by abiotic chemistry may lead is not yet clear. But it seems very likely that the first building blocks of nascent life were provided by amino acids and other small organic molecules such as are known to form readily in the laboratory and on celestial bodies. To what extent these substances arose on earth or were brought in by the falling comets and asteroids that contributed to the final accretion of our planet is still being debated.

The RNA World

Whatever the earliest events on the road to the first living cell, it is clear that at some point some of the large biological molecules found in modern cells must have emerged. Considerable debate in origin-of-life studies has revolved around which of the fundamental macromolecules came first—the original chicken-or-egg question.

The modern cell employs four major classes of biological molecules—nucleic acids, proteins, carbohydrates, and fats. The debate over the earliest biological molecules, however, has centered mainly on the nucleic acids, DNA and RNA, and the proteins. At one time or another, one of these molecular classes has seemed a likely starting point, but which? To answer that, we must look at the functions performed by each of these in existing organisms.

The proteins are the main structural and functional agents in the cell. Structural proteins serve to build all sorts of components inside the cell and around it. Catalytic proteins, or enzymes, carry out the thousands of chemical reactions that take place in any given cell, among them the synthesis of all other biological constituents (including DNA and RNA), the breakdown of foodstuffs, and the retrieval and consumption of energy. Regulatory proteins command the numerous interactions that govern the expression and replication of genes, the performance of enzymes, the interplay between cells and their environment, and many other manifestations. Through the action of proteins, cells and the organisms they form arise, develop, function, and evolve in a manner prescribed by their genes, as modulated by their surroundings.

The one thing proteins cannot do is replicate themselves. To be sure, they can, and do, facilitate the formation of bonds between their constituent amino acids. But they cannot do this without the information contained within the nucleic acids, DNA and RNA. In all modern organisms, DNA serves as the storage site of genetic information. The DNA contains, in encrypted form, the instructions for the manufacture of proteins. More specifically, encoded within DNA is the exact order in which amino acids, selected at each step from 20 distinct varieties, should be strung together to form all of the organism's proteins. In general, each gene contains the instructions for one protein.

DNA itself is formed by the linear assembly of a large number of units called nucleotides. There are four different kinds of nucleotides, designated by the initials of their constituent bases: A (adenine), G (guanine), C (cytosine), and T (thymine). The sequence of nucleotides determines the information content of the molecules, as does the sequence of letters in words.

Within all cells, DNA molecules are formed from two strands of DNA that spiral around each other in a formation called a double helix. The two strands are held together by bonds between the bases of each strand. Bonding is quite specific, so that A always bonds with T, and G is always partnered with C on the opposite DNA strand. This complementarity is crucial for faithful replication of the DNA strands prior to cell division.

During DNA replication, the DNA strands are separated, and each strand serves as a template for the replication of its complementary strand. Wherever A appears on the template, a T is added to the nascent strand. Or, if T is on the template, then A is added to the growing strand. The same is true for G and C pairs. In the characteristic double-helical structure of DNA, the two strands carry the same information in complementary versions, as do the positive and negative of the same photograph. Upon replication, the positive strand serves as template for the assembly of a new negative and the negative strand for that of a new positive, yielding two identical duplexes.

In order for DNA to fulfill its primary role of directing the construction of proteins, an intermediate molecule must be made. DNA does not directly participate in protein synthesis. That is the function of its very close chemical relative RNA.

Expression of DNA begins when an RNA molecule is constructed bearing the information for a gene contained on the DNA molecule. RNA, like DNA, is made up of nucleotides, but U (uracil) takes the place of T. Construction of the RNA molecule follows the same rules as DNA replication. The RNA copy, called a transcript, is a complementary copy of the DNA, with U (instead of T) inserted wherever A appears on the DNA template.

Most RNA transcripts, often after some modification, provide the information for the assembly of proteins. The sequence of nucleotides along the coding RNA, aptly called messenger RNA, specifies the sequence of amino acids in the corresponding protein molecule—three successive nucleotides (called a *codon*) in the RNA specify one amino acid to be used in the protein. The process is known as translation, and the correspondences between codons and amino acids define the genetic code.

Not all RNA molecules are messengers, however. Some of the RNAs participate in protein synthesis in other ways. Some actually make up the cellular machinery that constructs proteins. These are called ribosomal RNAs, and they may include the actual catalyst that joins amino acids by peptide bonds, according to the work of Harry Noller at the University of California at Santa Cruz. Other RNAs, called transfer RNAs, ferry the appropriate amino acids to the ribosome. As cell biology has progressed, even more functions for RNA have been discovered. For example, some RNA molecules participate in DNA replication, while others help process messenger RNAs.

Scientists considering the origins of biological molecules confronted a profound difficulty. In the modern cell, each of these molecules is dependent on the other two for either its manufacture or its function. DNA, for example, is merely a blueprint and cannot perform a single catalytic function, nor can it replicate on its own. Proteins, on the other hand, perform most of the catalytic functions but cannot be manufactured without the specifications encoded in DNA. One possible scenario for life's origins would have to include the possibility that two kinds of molecules evolved together, one informational and one catalytic. But this scenario is extremely complicated and highly unlikely.

The other possibility is that one of these molecules could itself perform multiple functions. Theorists consid-

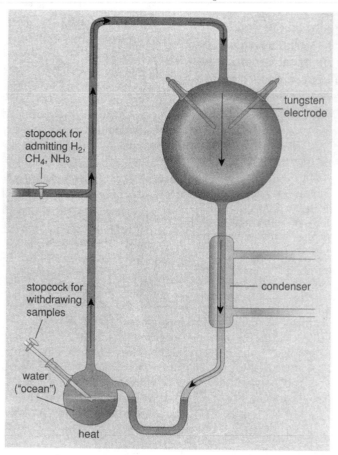

With this apparatus, Stanley Miller tried to simulate prebiotic storms in the laboratory of Harold Urey at the University of Chicago in the early 1950s. Repeated electrical discharges were passed through the "atmosphere," a mixture of hydrogen (H_2), methane (CH_4), ammonia (NH_3), and water vapor (H_2O) present in the upper sphere. Water (the "ocean") in the lower vessel, recycled by heating and condensation to simulate evaporation and rain, was found to contain increasing amounts of amino acids, the building blocks of proteins, and other potential biological compounds.

A question posed by origin-of-life studies revolves around the nature of the first biological macromolecules: Which, of DNA, RNA, or proteins, came first? In present-day organisms, DNA is the replicatable repository of genetic information. Its replication, mediated by complementary A-T and G-C pairing, is shown in the upper panel. Expression of the information stored in DNA, illustrated in the lower panel, takes place by way of transcription into RNA (also dependent on base pairing, with U, instead of T, pairing with A), followed by translation of the information from the resulting messenger RNA (mRNA) into polypeptides or proteins, which in turn determine the structural and functional properties of the cells. In protein synthesis, amino acids, carried by specific transfer RNA molecules (tRNA), are assembled into peptide chains on structures called ribosomes. The sequence of amino acids in the assembled peptide is dictated, according to the genetic code, by the sequence of nucleotides in the mRNA. The message is read by typical base pairing between triplets of nucleotides (codons) in the mRNA and complementary triplets (anticodons) in the tRNAs. Matching amino acids with tRNAs bearing the correct anticodon is ensured by the enzymes that catalyze the binding of the two kinds of molecules.

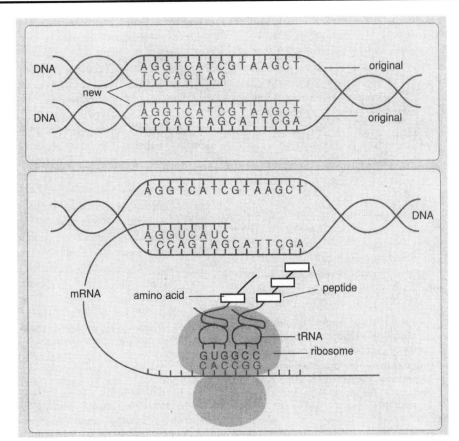

ering this possibility started to look seriously at RNA. For one thing, the molecule's ubiquity in modern cells suggests that it is a very ancient molecule. It also appears to be highly adaptable, participating in all of the processes relating to information processing within the cell. For a while, the only thing RNA did not seem capable of doing was catalyzing chemical reactions.

That view changed when in the late 1970s, Sydney Altman at Yale University and Thomas Cech at the University of Colorado at Boulder independently discovered RNA molecules that in fact could catalytically excise portions of themselves or of other RNA molecules. The chicken-or-egg conundrum of the origin of life seemed to fall away. It now appeared theoretically possible that an RNA molecule could have existed that naturally contained the sequence information for its reproduction through reciprocal base pairing and could also catalyze the synthesis of more like RNA strands.

In 1986, Harvard chemist Walter Gilbert coined the term "RNA world" to designate a hypothetical stage in the development of life in which "RNA molecules and cofactors [were]

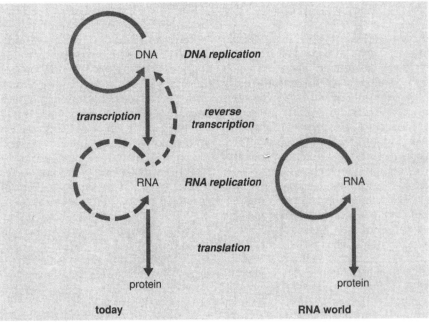

Information transfer in the modern cell progresses through the three cardinal steps shown in detail in the figure above. Added are two processes (*broken arrows*) found today only in cells infected by RNA viruses, which are reproduced either by direct replication (for example, the polio virus) or, via DNA, by reverse transcription followed by direct transcription (for example, the virus that causes AIDS). It is easily seen (*right*) that DNA is the most dispensable of the three kinds of macromolecules involved. It needs only RNA replication for RNA to be able to serve both as replicatable repository of the genetic information and as its translatable form. Passage to the present triad could then have taken place by reverse transcription of the information from RNA to DNA, associated with the development of DNA replication and transcription. This scenario is now accepted by most investigators in the field. As to the remaining RNA-protein diad, it is generally believed that RNA, which is replicatable and can exert some catalytic activities, preceded proteins, which are not replicatable. Such is the basis of the RNA-world model.

a sufficient set of enzymes to carry out all the chemical reactions necessary for the first cellular structures." Today it is almost a matter of dogma that the evolution of life did include a phase where RNA was the predominant biological macromolecule.

Origin and Evolution of the RNA World

As certain as many people are that the RNA world was a crucial phase in life's evolution, it cannot have been the first. Some form of abiotic chemistry must have existed before RNA came on the scene. For the purpose of this discussion, I shall call that earlier phase "protometabolism" to designate the set of unknown chemical reactions that generated the RNA world and sustained it throughout its existence (as opposed to metabolism—the set of reactions, catalyzed by protein enzymes, that support all living organisms today). By definition, protometabolism (which could have developed with time) was in charge until metabolism took over. Several stages may be distinguished in this transition.

In the first stage, a pathway had to develop that took raw organic material and turned it into RNA. The first building blocks of life had to be converted into the constituents of nucleotides, from which the nucleotides themselves had to be formed. From there, the nucleotides had to be strung together to produce the first RNA molecules. Efforts to reproduce these events in the laboratory have been only partly successful so far, which is understandable in view of the complexity of the chemistry involved. On the other hand, it is also surprising since these must have been sturdy reactions to sustain the RNA world for a long time. Contrary to what is sometimes intimated, the idea of a few RNA molecules coming together by some chance combination of circumstances and henceforth being reproduced and amplified by replication simply is not tenable. There could be no replication without a robust chemical underpinning continuing to provide the necessary materials and energy.

The development of RNA replication must have been the second stage in the evolution of the RNA world. The problem is not as simple as might appear at first glance. Attempts at engineering—with considerably more foresight and technical support than the prebiotic world could have enjoyed—an RNA molecule capable of catalyzing RNA replication have failed so far.

With the advent of RNA replication, Darwinian evolution was possible for the first time. Because of the inevitable copying mistakes, a number of variants of the original template molecules were formed. Some of these variants were replicated faster than others or proved more stable, thereby progressively crowding out less advantaged molecules. Eventually, a single molecular species, combining replicatability and stability in optimal fashion under prevailing conditions, became dominant. This, at the molecular level, is exactly the mechanism postulated by Darwin for the evolution of organisms: fortuitous variation, competition, selection, and amplification of the fittest entity. The scenario is not just a theoretical construct. It has been reenacted many times in the laboratory with the help of a viral replicating enzyme, first in 1967 by the late American biochemist Sol Spiegelman of Columbia University.

An intriguing possibility is that replication was itself a product of molecular selection. It seems very unlikely that protometabolism produced just the four bases found in RNA, A, U, G, and C, ready by some remarkable coincidence to engage in pairing and allow replication. Chemistry does not have this kind of foresight. In all likelihood, the four bases arose together with a number of other substances similarly constructed of one or more rings containing carbon and nitrogen. According to the present inventory, such substances could have included other members of the purine family (which includes A and G), pyrimidines (which include U, T, and C), nicotinamide and flavin, both of which actually engage in nucleotide-like combinations, and pterines, among other compounds.

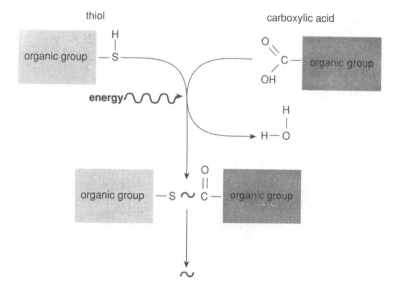

RNA world could not have been initiated or supported during the long time it took to evolve into the RNA-protein world without a strong, stable, and complex chemical underpinning, termed protometabolism, which the author argues must have followed chemical pathways similar to some of the pathways of present-day metabolism (congruence rule). At the center of protometabolism, the author envisages the thioester bond, a high-energy bond (symbolized by the red squiggle) that supports a large number of energy-requiring reactions in present-day metabolism and could have done the same in protometabolism. Thioesters are formed, with the help of energy, from the joining of thiols with carboxylic acids. (One molecule of water is expelled for each thioester bond formed.) The author calls the period in which thioester-bond energy drove protometabolism the thioester world.

The first nucleic acid-like molecules probably contained an assortment of these compounds. Molecules rich in A, U, G, and C then were progressively selected and amplified, once some rudimentary template-dependent synthetic mechanism allowing base pairing arose. RNA, as it exists today, may thus have been the first product of molecular selection.

A third stage in the evolution of the RNA world was the development of RNA-dependent protein synthesis. Most likely, the chemical machinery appeared first, as yet uninformed by genetic messages, as a result of interactions among certain RNA molecules, the precursors of future transfer, ribosomal, and messenger RNAs, and amino acids. Selection of the RNA molecules involved could conceivably be explained on the basis of molecular advantages, as just outlined. But for further evolution to take place, something more was needed. RNA molecules no longer had to be selected solely on the basis of what they *were*, but of what they *did*; that is, exerting some catalytic activity, most prominently making proteins. This implies that RNA molecules capable of participating in protein synthesis enjoyed a selective advantage, not because they were themselves easier to replicate or more stable, but because the proteins they were making favored their replication by some kind of indirect feedback loop.

This stage signals the limit of what could have happened in an unstructured soup. To evolve further, the system had to be partitioned into a large number of competing primitive cells, or protocells, capable of growing and of multiplying by division. This partitioning could have happened earlier. Nobody knows. But it could not have happened later. This condition implies that protometabolism also produced the materials needed for the assembly of the membranes surrounding the protocells. In today's world, these materials are complex proteins and fatty lipid molecules. They were probably simpler in the RNA world, though more elaborate than the undifferentiated "goo" or "scum" that is sometimes suggested.

Once the chemical machinery for protein synthesis was installed, information could enter the system, via interactions among certain RNA components of the machinery—the future messenger RNAs—and other, amino acid-carrying RNA molecules—the future transfer RNAs. Translation and the genetic code progressively developed concurrently during this stage, which presumably was driven by Darwinian competition among protocells endowed with different variants of the RNA molecules involved. Any RNA mutation that made the structures of useful proteins more closely dependent on the structures of replicatable RNAs, thereby increasing the replicatability of the useful proteins themselves, conferred some evolutionary advantage on the protocell concerned, which was allowed to compete more effectively for available resources and to grow and multiply faster than the others.

The RNA world entered the last stage in its evolution when translation had become sufficiently accurate to unambiguously link the sequences of individual proteins with the sequences of individual RNA genes. This is the situation that exists today (with DNA carrying the primary genetic information), except that present-day systems are enormously more accurate and elaborate than the first systems must have been. Most likely, the first RNA genes were very short, no longer than 70 to 100 nucleotides (the modern gene runs several thousand nucleotides), with the corresponding proteins (more like protein fragments, called peptides) containing no more than 20 to 30 amino acids.

It is during this stage that protein enzymes must have made their first appearance, emerging one by one as a result of some RNA gene mutation and endowing the mutant protocell with the ability to carry out a new chemical reaction or to improve an existing reaction. The improvements would enable the protocell to grow and multiply more efficiently than other protocells in which the mutations had not appeared. This type of Darwinian selection must have taken place a great many times in succession to allow enzyme-dependent metabolism to progressively replace protometabolism.

The appearance of DNA signaled a further refinement in the cell's information-processing system, although the date of this development cannot

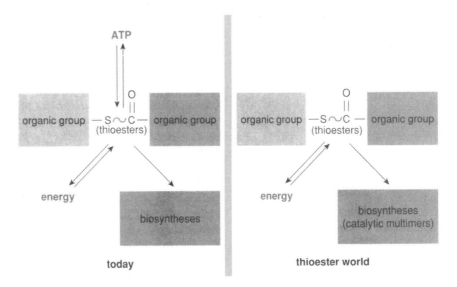

In present-day organisms ATP fuels many energy-requiring processes by way of thioesters (*left*). Conversely, a number of energy-yielding reactions support the assembly of ATP by way of thioesters (other mechanisms of ATP assembly also exist but require complex membrane-embedded systems not likely to have been present in the early prebiotic world). In prebiotic days, thioesters could have supported protometabolism before ATP came on the scene (*right*). In this phase, biosynthetic products include multimers assumed to have served as catalysts in the thioester world. Compare the dispensability of ATP in the thioester world to that of DNA in the RNA world.

be fixed precisely. It is not even clear whether DNA appeared during the RNA world or later. Certainly, as the genetic systems became more complex, there were greater advantages to storing the genetic information in a separate molecule. The chemical mutations required to derive DNA from RNA are fairly trivial. And it is conceivable that an RNA-replicating enzyme could have been co-opted to transfer information from RNA to DNA. If this happened during the RNA world, it probably did so near the end, after most of the RNA-dependent machineries had been installed.

What can we conclude from this scenario, which, though purely hypothetical, depicts in logical succession the events that must have taken place if we accept the RNA-world hypothesis? And what, if anything, can we infer about the protometabolism that must have preceded it? I can see three properties.

First, protometabolism involved a *stable* set of reactions capable not only of generating the RNA world, but also of sustaining it for the obviously long time it took for the development of RNA replication, protein synthesis, and translation, as well as the inauguration of enzymes and metabolism.

Second, protometabolism involved a *complex* set of reactions capable of building RNA molecules and their constituents, proteins, membrane components, and possibly a variety of coenzymes, often mentioned as parts of the catalytic armamentarium of the RNA world.

Finally, protometabolism must have been *congruent* with present-day metabolism; that is, it must have followed pathways similar to those of present-day metabolism, even if it did not use exactly the same materials or reactions. Many abiotic-chemistry experts disagree with this view, which, however, I see as enforced by the sequential manner in which the enzyme catalysts of metabolism must have arisen and been adopted. In order to be useful and confer a selective advantage to the mutant protocell involved, each new enzyme must have found one or more substances on which to act and an outlet for its prod-

uct or products. In other words, the reaction it catalyzed must have fitted into the protometabolic network. To be sure, as more enzymes were added and started to build their own network, new pathways could have developed, but only as extensions of what was initially a congruent network.

The Thioester World

It may well be, then, that clues to the nature of that early protometabolism exist within modern metabolism. Several proposals of this kind have been made. Mine centers around the bond between sulfur and a carbon-containing entity called an acyl group, which yields a compound called a thioester. I view the thioester bond as primeval in the development of life. Let me first briefly state my reasons.

A thioester forms when a thiol (whose general form is written as an organic group, R, bonded with sulfur and hydrogen, hence R-SH) joins with a carboxylic acid (R'-COOH). A molecule of water (H_2O) is released in the process, and what remains is a thioester: R-S-CO-R'. The appeal in this bond is that, first, its ingredients are likely components of the prebiotic soup. Amino acids and other carboxylic acids are the most conspicuous substances found both in Miller's flasks and in meteorites. On the other hand, thiols may be expected to arise readily in the kind of volcanic setting, rich in hydrogen sulfide (H_2S), likely to have been found on the prebiotic earth. Joining these constituents into thioesters would have required energy. There are several possible mechanisms for this, which I shall address later. For the time being, let us assume thioesters were present. What could they have done?

The thioester bond is what biochemists call a high-energy bond, equivalent to the phosphate bonds in adenosine triphosphate (ATP), which is the main supplier of energy in all living organisms. It consists of adenosine monophosphate (AMP)—actually one of the four nucleotides of which RNA is made—to which two phosphate groups are attached. Splitting either of these two phosphate bonds in ATP generates energy, which fuels

the vast majority of biological energy-requiring phenomena. In turn, ATP must be regenerated for work to continue.

It is revealing that thioesters are obligatory intermediates in several key processes in which ATP is either used or regenerated. Thioesters are involved in the synthesis of all esters, including those found in complex lipids. They also participate in the synthesis of a number of other cellular components, including peptides, fatty acids, sterols, terpenes, porphyrins, and others. In addition, thioesters are formed as key intermediates in several particularly ancient processes that result in the assembly of ATP. In both these instances, the thioester is closer than ATP to the process that uses or yields energy. In other words, thioesters could have actually played the role of ATP in a thioester world initially devoid of ATP. Eventually, their thioesters could have served to usher in ATP through its ability to support the formation of bonds between phosphate groups.

Among the substances that form from thioesters in present-day organisms are a number of bacterial peptides made of as many as 10 or more amino acids. This was discovered by the late German-American biochemist Fritz Lipmann, the "father of bioenergetics," toward the end of the 1960s. But even before that, Theodor Wieland of Germany had found in 1951 that peptides form spontaneously from the thioesters of amino acids in aqueous solution.

The same reaction could be expected to happen in a thioester world, where amino acids were present in the form of thioesters. Among the resulting peptides and analogous multi-unit macromolecules, which I like to call multimers to emphasize their chemical heterogeneity, a number of molecules could have been structurally and functionally similar to the small catalytic proteins that inaugurated metabolism. I therefore suggest that multimers derived from thioesters provided the first enzyme-like catalysts for protometabolism.

The thioester world thus represents a hypothetical early stage in the devel-

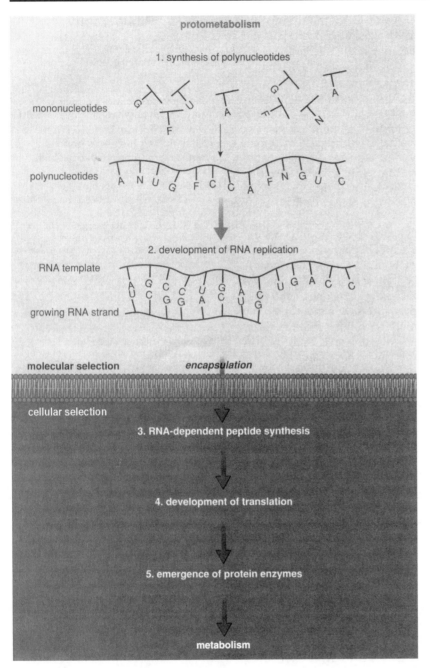

protometabolism

1. synthesis of polynucleotides

mononucleotides

polynucleotides

2. development of RNA replication

RNA template

growing RNA strand

molecular selection   *encapsulation*

cellular selection

3. RNA-dependent peptide synthesis

4. development of translation

5. emergence of protein enzymes

metabolism

Early evolution from protometabolism to metabolism presumably took place through the five main steps represented here. In a first stage, simple building blocks, of the kind found in meteorites or reproduced in simulation experiments, were converted into mononucleotides (including some not found in RNA), which were further assembled into polynucleotides. With the development of replication in the second step, typical RNA molecules were selected out, in what may have been the first instance of Darwinian competition at the molecular level. In a third step, RNA molecules must have interacted together and with amino acids to build the first peptides. At this stage or earlier, these prebiotic systems must have become encapsulated by primitive fatty membranes, forming primitive cells, or protocells. Darwinian competition could then take place among protocells, instead of among RNA molecules. The first peptide-synthesizing machinery most likely joined amino acids more or less at random. Translation and the genetic code emerged later through another complex set of molecular interactions, subject to competition and selection. Once systems of RNA replication and translation of sufficient fidelity were installed, typical Darwinian evolution by mutations of RNA genes and competition among the resulting variant protocells could take place. It is suggested that enzymes arose, one by one, in this manner. The advantages conferred by each new enzyme allowed mutant protocells to multiply faster than the others. As peptide enzymes emerged and assumed their functions, metabolism gradually replaced protometabolism.

opment of life that could have provided the energetic and catalytic framework of the protometabolic set of primitive chemical reactions that led from the first building blocks of life to the RNA world and subsequently sustained the RNA world until metabolism took over.

This hypothesis implies that thioesters could form spontaneously on the prebiotic earth. Assembly from thiols and acids could have occurred, although in very low yield, in a hot, acidic medium. They could also have formed in the absence of water, for example, in the atmosphere. Perhaps a more likely possibility is that thioesters formed, as they do in the present world, by reactions coupled to some energy-yielding process. The American chemist Arthur Weber, formerly of the Salk Institute, now at the NASA Ames Research Center in California, has described several simple mechanisms of this sort that could have operated under primitive-earth conditions.

So far, these ideas are highly speculative, being supported largely by the need for congruence between protometabolism and metabolism, by the key—and probably ancient—roles played by thioesters in present-day metabolism, and by the likely presence of thioesters on the prebiotic earth. But some experimental evidence has been obtained that supports the thioester world model.

I have already mentioned the work of Wieland, Lipmann, and Weber. Recently, highly suggestive evidence has come from the laboratory of Miller, where researchers have obtained under plausible prebiotic conditions the three molecules—cysteamine, β-alanine, and pantoic acid—that make up a natural substance known as pantetheine. They have also observed the ready formation of this compound from its three building blocks under prebiotic conditions. It so happens that pantetheine is the most important biological thiol, a catalytic participant in a vast majority of the reactions involving thioester bonds.

A Cosmic Imperative

I have tried here to review some of the facts and ideas that are being considered to account for the early stages in the spontaneous emergence of life on earth. How much of the hypothetical mechanisms considered will stand the test of time is not known. But one affirmation can safely be made, regardless of the actual nature of the processes that generated life. These processes must have been highly *deterministic*. In other words, these processes were inevitable under the

---

. . . life is an obligatory manifestation of matter, bound to arise where conditions are appropriate.

---

conditions that existed on the prebiotic earth. Furthermore, these processes are bound to occur similarly wherever and whenever similar conditions obtain. This must be so because the processes are chemical and are therefore ruled by the deterministic laws that govern chemical reactions and make them reproducible.

It also seems likely that life would arise anywhere similar conditions are found because many successive steps are involved. A single, freak, highly improbable event can conceivably happen. Many highly improbable events—drawing a winning lottery number or the distribution of playing cards in a hand of bridge—happen all the time. But a string of improbable events—the same lottery number being drawn twice, or the same bridge hand being dealt twice in a row—does not happen naturally.

All of which leads me to conclude that life is an obligatory manifestation of matter, bound to arise where conditions are appropriate. Unfortunately, available technology does not allow us to find out how many sites offer appropriate conditions in our galaxy, let alone in the universe. According to most experts who have considered the problem—notably, in relation with the Search for Extraterrestrial Intelligence project—there should be plenty of such sites, perhaps as many as one million per galaxy. If these experts are right, and if I am correct, there must be about as many foci of life in the universe. Life is a cosmic imperative. The universe is awash with life.

Bibliography

Deamer, D. W., and G. L. Fleischaker, eds. 1994. *Origins of Life: The Central Concepts.* Boston: Jones and Bartlett.

de Duve, C. 1991. *Blueprint for a Cell: The Nature and Origin of Life.* Burlington, N.C.: Neil Patterson Publishers, Carolina Biological Supply Company

de Duve, C. 1995. *Vital Dust: Life as a Cosmic Imperative.* New York: Basic Books.

Gilbert, W. 1986. The RNA World. *Nature* 319: 618.

Gilbert, W. 1987. The exon theory of genes. *Cold Spring Harbor Symposium on Quantative Biology* 52: 901-905.

Keefe, A. D., G. L. Newton and S. L. Miller. 1995. A possible prebiotic synthesis of pantetheine, a precursor of coenzyme A. *Nature* 373: 683-685.

Miller, S. L. 1953. A production of amino acids under possible primitive earth conditions. *Science* 117: 528-529.

Noller H. F, V. Hoffarth and L. Zimniak. 1992. Unusual resistance of peptidyl transferase to protein extraction procedures. *Science* 256: 1416-1419.

Schopf, J. W. 1992. The oldest fossils and what they mean. In *Major Events in the History of Life,* ed. J. W. Schopf. Boston: Jones and Bartlett, pp. 29-63.

Schopf, J. W. 1993. Microfossils of the early Archaean apex chert: new evidence of the antiquity of life. *Science* 260: 640-646.

THOUGHT EVOKERS

- What does the term "organic" chemistry mean?

- What was the first proof that a special "vital force" is not needed for organic syntheses? Explain.

- How do scientists postulate that DNA and RNA were involved in the early evolution of life on earth?

12

The Rape of the Oceans

**America's last frontier is
seriously overfished,
badly polluted,
poorly managed,
and in deepening trouble**

By Michael Satchell

With Capt. Joe Testaverde at the helm, the trawler Nina T slips into its Gloucester, Mass., mooring at sunset. Joe's father, Salvatore Testaverde, and his Sicilian father before him were Gloucester fishermen, and the family has trawled Northeastern waters for close to 80 years. This night seems a comforting continuum as the fish are unloaded and the crew members josh with dockside onlookers, exaggerating the size of the catch. The scene in this snug New England harbor is as timeless and reassuring as the tides—and as deceptive as a roseate dawn.

In bygone years, Joe Testaverde's father and grandfather would return to port with their boats packed from bilge to gunwale with haddock and flounder, and with jumbo codfish weighing 50 pounds or better. Sal Testaverde recalls pulling up 5,000 pounds of cod in a one-hour tow. Today, if he can find them, son Joe might haul in 2,000 pounds of middling-sized cod in eight hours of hard trawling. And he won't even waste time searching for flounder and haddock. The Nina T, more than likely, will return with hake, whiting, spiny dogfish, or skate—species despised in Sal Testaverde's day as "trash" fish. Shipped abroad or retailed in ethnic markets for about $1 a pound, they are the dominant and devalued currency of the Georges Bank, once the nation's richest fishing ground.

The precipitous, perhaps irreversible decline of New England's groundfish is one of the major casualties of an unrelenting assault on the nation's coastal oceans. The principal problems are overfishing, burgeoning seaside development, loss of coastal wetlands, and pollution of bay and estuary fish breeding grounds. Compounding these pressures is the profligate waste of hundreds of millions of pounds of edible "bycatch" fish. And a political consensus between fishing and federal bureaucrats to better manage the vast and valuable marine resources has yet to be reached. This looming disaster extends from the coastlines out to the 200-mile limit in the Atlantic and Pacific oceans, the Gulf of Mexico, the Gulf of Alaska, and the Bering Sea. Together, they represent the country's last open frontier.

About 1.5 million dolphins and other small whales are killed each year by tuna fishermen, by pollution, or in targeted hunts.

These 2.2 million square miles contain about one fifth of the world's harvestable seafood; enormous populations of marine birds and mammals; and spectacular undersea reefs, banks, and gardens teeming with life forms that have barely been studied. "We have two choices—conserve and develop a sustainable resource, or squander and destroy it," says Roger McManus, head of the Center for Marine Conservation, the only national environmental group devoted solely to the oceans' welfare. "Our record so far is abysmal."

Earth Summit. There is a growing belief among environmentalists that the world's overexploited and ailing oceans will replace tropical rain forests as the next global ecology concern. At the United Nations Earth Summit in Rio de Janeiro that wrapped up last week, participants pledged to try and control overfishing, pollution, and coastal development. But the agreement contained no bold new initiatives and few specific goals or enforcement mechanisms. Not surprisingly, amid the rancorous parley that saw the United States excoriated for its independent stance on global warming, forest protection, and biodiversity, marine issues drew scant attention from the media or from official delegates.

But elsewhere, the alarms are sounding—especially in America. Front-line U.S. environmental groups like Greenpeace, the National Audubon Society, and the World Wildlife Fund are following the lead of the Center for Marine Conservation and turning their attention to ocean biodiversity. Until recently, they paid little attention to such unglamorous issues as fish. They favored instead emotional, hot-button topics like the killing of dolphins, harp seals, and whales; birds and mammals dying in drift nets; or saving endangered species like manatees and sea turtles. "Marine fisheries are the nation's single most threatened resource," says Amos Eno of the nonprofit National Fish and Wildlife Foundation. "There has been weak involvement by major environmental groups and poor federal management."

Close to half of U.S. coastal finfish stocks are now overexploited—meaning that more are being caught than are replenished by natural reproduction. Scientists say 14 of the most valuable species—New England groundfish, red snapper, swordfish, striped bass, and Atlantic bluefin tuna among them—are threatened with commercial extinction, meaning that too few would remain to justify the cost of catching them.

Only drastic conservation measures will restore these threatened stocks. However, a five-to-l0-year fishing ban to allow rebuilding could spell economic disaster for segments of the fishing industry. Virtually all the remaining commercial finfish stocks—except in Alaskan waters—are now being harvested to their limits.

Further fishing pressure could put more species in jeopardy. Between 1986 and 1991, the finfish and shellfish catch off the lower 48 states declined by 50 million pounds, from 4.8 billion pounds to 4.3 billion. The harvest of menhaden—used to make poultry feed, fish oil, and other products—dropped last year by 300 million pounds. "When Nature is at her best, you can fish with impunity and get away with it," says Lee Weddig, executive director of the National Fisheries Institute, a trade organization. "We can do that no longer."

Pollution's toll. The assault on the oceans begins at the shoreline. About half the U.S. population lives within 50 miles of the coastline, and the booming development is hard to control. The result: massive changes in coastal ecology that are destroying or damaging habitat for finfish and shellfish. Coastal wetlands are gobbled up: Louisiana, for example, loses about 50 square miles of piscatorial breeding ground annually. In California, only 9

In 1990, Japanese drift netters dumped 39 million unwanted fish, 700,000 sharks, 270,000 sea birds, and 26,000 mammals. Most were dead.

---

percent of the state's original 3.5 million acres of coastal wetlands remain. These bays and estuaries are the breeding grounds and nurseries for fully 75 percent of commercial-seafood species. But they are being increasingly befouled by sewage, industrial waste water, and runoff from cities and farms.

Half the fish in areas polluted by toxic chemicals fail to spawn and suffer from weakened immune systems. Chemical nutrients from smokestacks and sewers and from pesticides and fertilizers used on farms and front lawns stimulate explosive algae growth that blocks sunlight and eventually depletes the waters of oxygen, creating undersea dead zones. One

---

The world eats 90 million metric tons of seafood annually. Replacing it with red meat would require 200 million steers or 750 million hogs.

---

acre of Narragansett Bay, R.I., or Delaware Bay, for example, now gets more nitrogen and phosphorus annually from urban and farm runoff than an average acre of cotton, soybeans, or wheat grown in the United States.

At any given time, fully one third of the nation's oyster, clam, and other shellfish beds are closed because of contamination. Some 27 marine mammals and birds in American coastal waters are now listed as threatened or endangered, and the rising phenomenon of mass die-offs of dolphins and seals is blamed on toxins like PCBs that are rapidly accumulating in the marine environment. And as every beachgoer knows, trash is piling up in ever increasing amounts. Some shorelines along the Gulf of Mexico are strewn with 2 tons of marine debris per mile.

This assault on the marine environment is being felt most keenly by the nation's commercial fisheries. Both the fiercely independent industry and the National Marine Fisheries Service, which tries hard to regulate it, are struggling over ways to control overfishing, reduce the number of boats, curb the waste of bycatch fish, and find more efficient ways to manage and harvest the stocks. At present, fisheries are the least regulated of public resources. Enforcement of even the tax regulations is spotty, compliance is weak, and there is simply no real incentive for individual fishermen to conserve.

The severely depleted finfish stocks in the lower 48 states, and the government's failure to manage effectively the coastal oceans for the commonweal, raise several questions. Should the seas up to the 200-mile limit be treated as a public resource—like federal lands—and administered like other natural assets? Industries pay the government to drill for offshore oil, cut national-forest timber, and mine coal on public lands. Should fishermen pay to harvest seafood in federal waters? Before the Taylor Grazing Act of 1934, Western rangelands were nearly destroyed by cattle because ranchers had free access. Could this classic tragedy-of-the-commons be repeated in America's oceans?

Questions of control. The root cause of today's overfishing goes back to 1977, when the Magnuson Act extended U.S. coastal jurisdiction from 12 miles to 200 miles and the dominant foreign factory vessels were kicked

---

Atlantic bluefin tuna—a sushi delicacy—have declined 90 percent since 1970. In Japan, a single fish can wholesale for $30,000.

---

out of American waters. Ironically, the nation managed to gain control of its coastal oceans but in so doing, simply traded overfishing by foreigners for unrestrained plunder by domestic fishermen.

Meanwhile, the rust-bucket American fleet was gradually replaced—thanks to more than $500 million in federal loan guarantees—with efficient, high-tech boats. And a $1.3 billion fleet of some 70 factory vessels, American-registered but mostly owned by Japanese, Scandinavian, and Korean interests, now plies the Gulf of Alaska and the Bering Sea. From Gloucester Mass., to Cordova, Alaska, the industry has too many fishing boats—but there is no consensus on how to control this armada or limit access to the fisheries.

The fishing capacity of the new boats is staggering and, more than anything else, has exacerbated the destructive pressure on the fish stocks. Just a decade ago, many fishermen still used binoculars, shoreside

---

PCBs in the milk of Canadian Beluga whales—"The world's most polluted animals"—are 3,400 times the safe levels for drinking water.

---

landmarks, and oilcan buoys to mark and relocate productive areas. Today, the smallest vessels employ sophisticated electronics that can pinpoint a single codfish at 100 fathoms or guide a captain to within 100 feet of a favorite hot spot. East Coast fishermen seeking increasingly scarce swordfish in the warm Gulf Stream waters, for example, once lowered thermometers over the side or scanned the surface for blue warm-

water eddies. Now they can get ocean temperatures faxed to their boats via satellite.

To marine biologists, the decline of New England's prized cod, flounder, and haddock stocks is a warning for all American fisheries. Many blame the mess on weak leadership by the National Marine Fisheries Service and on the New England Fishery Management Council, one of eight regional groups set up nationwide by the NMFS to govern the industry. Critics contend that the councils are dominated by commercial-fishing interests that are less concerned with conservation than with maximizing profits. In 1981, for example, New England fishermen pressured their regional council to remove all catch quotas on cod, haddock, and flounder. That led to the disastrous depletion of the most valuable commercial species. Short of a politically difficult multiyear fishing freeze, the NMFS wants

to limit gradually the fishermen's days at sea, increase net-mesh sizes to reduce mortality of small, unwanted fish, and ban new boats from entering the fishery. The alternative Darwinian solution is to let the fishermen fight for the dwindling supply.

David versus Goliath. Rooted in the colonial era and steeped in tradition, New England fisheries have long resisted regulation. Alaska's modern fisheries, developed in recent decades, more readily accept tight controls. But common problems are magnified in the Northern Pacific and they threaten the resource. Alaska's high-tech, industrial-strength fleet has too many boats, and there is enormous waste.

Some 70 factory ships are locked in fierce competition with thousands of small vessels in the Bering Sea and Gulf of Alaska. Together, the David and Goliath fleets have enough capacity to scoop up the entire annual Alaskan quota of about 4.8 billion

---

Can sharks survive?

After 400 million years of evolution, the shark is the top predator in the ocean. At the apex of the food chain, cruising its domain in perpetual, primordial motion, it is the undisputed king of the under-sea jungle.

But it has taken man—an even more ruthlessly efficient killer—just a single decade to threaten the shark's survival. Hunted for their meat and their fins, many species are in a dizzying ecological plunge. "In America, and around the world, sharks are being fished to oblivion," says University of Miami shark expert Samuel Gruber. "Without drastic conservation measures, some species will be lost."

Making soup

Sharks became gourmet fare in the late 1970s, after the National Marine Fisheries Service encouraged the harvesting of what had been widely regarded as an underutilized "trash" fish—good only for crab bait or sport angling. But the heaviest pressure to harvest sharks now comes from the rising demand in Asia for shark fins to make soup. Depending on the species, fins can be worth $5 to $30 a pound to U.S. fishermen. In parts of Asia, choice fins sell for as much as $150 a pound. The boom has boosted the number of commercial boats targeting sharks or taking them as a bonus bycatch. It has also led to the gruesome and wasteful practice of "finning": fishermen cut off the two valuable fins and toss the helpless animal back to starve.

The prospects for saving endangered shark species are complicated by the fact that the animals are slow to mature and reproduce. But for marine biologist Gruber, the idea of seas lacking sharks is deeply disturbing. Without the primary predator, fish populations would rapidly expand, possibly setting off a chain reaction that

could upset the delicately balanced marine ecosystem all the way down to plankton. "We don't know enough about the true mechanics of the biosphere to predict with certainty what would happen," he says. "But we do know a diverse, balanced ecosystem is good. And for that, we need sharks."

Medical value

Some 350 species—from 8-inch cigar sharks to 35-foot whale sharks—inhabit the oceans. Beyond their food value, they show promising medical potential for humans. Their cartilage is used as artificial skin for burn victims, their corneas have been used for human replacement, and shark liver oil is a principal ingredient in many hemorrhoid ointments. Scientists are interested in studying their high resistance to cancers and their regenerative powers that allow wounds to heal rapidly. "It's amazing, the damage they suffer and how quickly they recover," Gruber says. "I've seen them with badly lacerated corneas and massive wounds from mating that heal very quickly. I've seen them with stingray spines stuck through their mouths or piercing the brain or heart cavity. They do fine."

A federal shark-recovery plan announced in January calls for an end to finning and for strict commercial- and sport-fishing limits designed to protect 39 coastal and high-seas species whose survival is threatened. Gruber says the plan is a good beginning, although for him, it comes about four years too late. In 1988, he was forced to abandon 30 years of behavioral studies of lemon sharks in the Florida Keys after the population was wiped out for crab bait. "When I was a student, I thought the oceans were so wide and so vast that it was impossible to degrade the system," he says. "I thought it was a great sink for chemicals and pollution, and there was no way to put a dent in the marine populations. Boy, was I wrong."

pounds of groundfish in less than six months. With factory ships worth $75 million tied up in port and shoreside plants served by the small vessels periodically idled, with no fish to process, there is relentless lobbying by the rival fleets for a bigger share of the catch.

The very fecundity of the Alaskan waters, like that of other regions, encourages waste. Fishing everywhere produces bycatch that is thrown back—invariably dead or dying—because the fish are too small, too big, have lower market value than the target species, or because it is illegal to keep them. For every pound of shrimp hauled from the Atlantic Ocean and the Gulf of Mexico, for example, an estimated 9 pounds of red snapper, croaker, mackerel, sea trout, spot, drum, and other species are brought up in the nets and tossed overboard. This annual shrimp-fishery bycatch alone is estimated by the NMFS at more than 1 billion pounds—a waste that equals 10 percent of the entire U.S. harvest.

"National scandal." In 1990, Alaskan trawlers fishing for pollock and cod jettisoned some 25 million pounds of halibut—worth about $30 million— plus vast quantities of salmon and king crab, because they were an incidental, prohibited bycatch. They also reported throwing away 500 million pounds of ground-fish because they were the wrong size or to save space for more valuable species.

NMFS officials think the waste is actually higher than 550 million pounds because that total is largely reported by vessel captains and thus is unverifiable. Larry Cotter, a Juneau, Alaska, fisheries consultant and former bycatch chairman for the NMFS regional management council, calls the waste "a national scandal and an unconscionable disgrace."

Another pressing problem in Alaskan waters emanates from the traditional free-for-all culture that allows anyone with a boat to try his luck. So many fishermen compete for the valuable Pacific halibut that the entire year's quota—once harvested over a leisurely six-month season—is now taken in two frantic 24-hour periods.

Last year's derby for the giant flat-fish was typical. At noon on September 3, some 6,000 vessels raced out of Alaskan ports in a maritime version of the Oklahoma land rush and began hauling in halibut like sharks in a feeding frenzy. At noon the next day, when the season ended, many vessels limped back to port dangerously overloaded. The Coast Guard handled 29 Mayday calls from sinking boats. There was no loss of life, but during the earlier one-day spring season, two crewmen had drowned. The September boats landed 23.7 million pounds, but about 4 million pounds rotted because it couldn't be frozen quickly enough. And the massive glut hitting the market in one day meant no fresh halibut for most consumers.

To fix many of the problems on all three coasts, the National Marine Fisheries Service is eager to manage the nation's seafood stocks through a system of individual transferable quotas, called ITQs. Based on their previous fishing history, boat owners would be given a permit allowing them to harvest a fixed amount of finfish or shellfish each year. The permits could be leased, sold, or passed on in a family. "We need to get fishermen to act more like farmers—give them an ownership privilege and a vested interest in the stocks," says NMFS conservation chief Richard Schaefer. "Fishermen don't own a fish until it's flopping around on the deck. If they own it before it is caught, they will manage it rationally."

The ITQ system is already being tested in two East Coast clam fisheries, and Alaskan halibut is one of several other species under considera-

tion for ITQs. But while many fishermen favor temporary limits on the number of boats entering the industry, there is little enthusiasm for permits, quotas, or other regulations. Their reluctance exemplifies the freewheeling, entrepreneurial independence that fisherman cling to—the right to get rich or go broke, unfettered or unprotected by Uncle Sam. "ITQs will destroy us," says Edward Lima, executive director of the Cape Ann Vessel Association in Gloucester. "Corporations will buy up permits and swallow us like corporate farms gobbling up family farmers." His friend Joe Testaverde agrees. And he objects on less tangible grounds, as well. "Quotas remove the lure and romance of fishing," he says. "Part of the thrill is to go out one day and catch nothing, then come home the next and beat out the other guy."

But dwindling stocks and the urgent need for conservation mean that fishermen will have to adapt and change—or risk destroying the very resource that sustains them. Like any good fisherman, Joe Testaverde is cheerfully optimistic. If the cod, haddock, and flounder don't come back, he will fish for something else. "Yesterday, on the sonar screens, I saw tons and tons of small fish—whiting, mackerel, butterfish, squid, and herring," he says. "If I didn't see that, I'd be scared of the future. We're the last surviving hunters and gatherers. I know I'm *always* gonna catch fish." Testaverde's attitude reflects much of what is right—and what is wrong— with the fishing industry. It is also the reason why America's oceans are in deepening trouble.

THOUGHT EVOKERS

- Why is it a growing belief that the condition of the world's oceans may soon become our next global concern?

- How does biodiversity relate to the stability of an ecosystem?

- Describe the major events that are contributing to the alleged destruction of the world's marine fishery.

13

SEAGOING SPACESHIPS

EXPLORING THE MARINE WORLD BY THE LIGHT OF SEA SPRITES CALLED LARVACEANS

By Cheryl Lyn Dybas

The object looming on the video screen looks like a ghostly underwater spaceship slowly revolving through the ocean depths. In fact, it's a "winged" animal called a larvacean—a web-encased, tadpole-like creature that scientists have ventured miles out to sea to study from the *Point Lobos,* the Monterey Bay Aquarium Research Institute's 110-foot research vessel. Marine biologist Mary Silver, of the University of California at Santa Cruz, and other researchers are using a Volkswagen-sized remotely operated vehicle (ROV), which is tethered with bright yellow fiber-optic cable to the *Point Lobos,* to learn more about the importance of larvaceans in ocean ecology.

Superficially resembling jellyfish, larvaceans are inch-long, finger-shaped animals that spin fragile webs of mucus around themselves. Their gossamer "houses" are about the size of a walnut or smaller, but in certain species can reach several feet in width. Like fine nets, the webs act as food filters to trap microscopic drifting plants of the sea called phytoplankton.

There are some 70 species of larvaceans, all belonging to a group of animals called tunicates, which include the more familiar sea squirts. Unlike most of the tunicates, adult larvaceans do not metamorphose but remain in the tadpole-shaped larval stage, hence the name *larvacean*. The three main larvacean families are Oikopleuridae, which spin bubble-shaped houses; Fritillaridae, with

domains shaped like butterflies; and Kowalevskiidae, rare creatures with houses that resemble umbrellas.

It is a chilly morning in mid-February, and the biologists aboard the *Point Lobos* have chosen a location several miles out in Monterey Bay to look for one of the most beautiful of the larvaceans. Mary Silver speaks into her communications headset: "We're at 250 meters [more than 700 feet below the surface], and we've 'made contact.'"

A butterfly-shaped *Fritillaria* floats on the screen in the darkened control room below deck. Images are relayed via television link from a high-resolution camera mounted on the front of the ROV, named Ventana, Spanish for "window." The *Point Lobos* pitches and rolls, groaning in the rough bay waters, but the cathedral stillness of the dim world hundreds of feet below is broken only by the whirr of Ventana's thrusters.

> **Divers have compared floating among these chiffon creatures to being brushed by thousands of fairy wings.**

Movable propellers facing in different directions allow the ROV to be maneuvered with sufficient precision to study the delicate organisms. "In the past, we used nets," says Silver, "but by the time we got the larvaceans to the surface, their fragile bodies had

collapsed. Now we can observe them intact and in their natural environment through Ventana's camera."

Thanks to ROVs and various technological advances, scientists are discovering that larvaceans and other small creatures are more numerous and widespread in the oceans than they'd previously believed. "You need an undersea vehicle to really study these animals," explains Silver, "and not many places in the ocean have been explored using one. Just about everywhere we have been able to look, however, we've found larvaceans."

The number of larvaceans drifting in the sea are almost too great to imagine. Where these wraiths are abundant, they often reach a density of several thousand per cubic meter of seawater. Along the Atlantic Coast, in Delaware Bay, one sampling yielded more than 270,000 larvaceans. *Oikopleura dioica,* for example, occurs there in numbers of 3,000 to 5,000 per cubic meter.

In July 1968, a shimmering pink light glowed in the waters of Saanich Inlet on Vancouver Island, British Columbia, the result of a concentration of *O. dioica.* By mid-summer, this larvacean's numbers had reached the millions, turning the inlet a luminescent rose. Divers have compared floating among these chiffon creatures to being brushed by thousands of fairy wings.

According to marine biologist Alice Alldredge, of the University of California at Santa Barbara, the larvacean web is one of the most complex external structures built by any organism. Continuous beating of the animal's tail draws water into its house, which is interlaced with a variety of filters that exclude particles too large to ingest. Water that has been sieved for phytoplankton flows up the arched "wings" and back out another set of openings into the sea. Every four hours or so—the length of time it takes for the filters to become clogged with debris—the larvacean discards its house and secretes a new one. Sometimes the accumulated material becomes so heavy that the web independently pulls free of the animal, compresses, and twists into ropelike strands that sink rapidly. When this happens, new houses are manufactured in a few minutes.

The life expectancy of a house depends at least partly upon the larvacean's interactions with other creatures. Large, strong swimmers such as

> **"Larvaceans are like floating food islands; they sustain all kinds of small animals that have hooked appendages for grazing."**

squid and blue sharks may destroy the delicate webs with their passing wakes. When larvaceans are trapped in a school of hake or other fish, their filmy houses are often torn asunder.

A fascination with the feeding nets of larvaceans led marine scientists to the discovery, in the 1800s, of nanoplankton, which are extremely small, drifting sea plants. These botanical Lilliputians were first observed by German biologist Hans Lohmann, who found them in the filters of larvacean houses. British scientist Sir Alister Hardy wrote in his 1965 classic on marine biology, *The Open Sea: Its Natural History*, "Man has not yet succeeded in devising suitable means for the capture of the tiny members of the plankton. That larvaceans should have solved the prob-

lem so efficiently makes one marvel all the more at these unusual animals."

The larvacean filters trap the small organisms that make up nanoplankton, such as dinoflagellates and coccoliths. "Few animals but larvaceans are able to capture and feed on these dwarfs," adds Silver.

Larvaceans act as a vital link between the tiny plankton they feed on and larger animals in the oceanic food chain, which in turn feed upon them. Predators of larvaceans include young herring, sardines, and flounders. Juvenile plaice, or flatfish, may consume as many as 30 larvaceans a day. Siphonophores, relatives of the Portuguese man-of-war jellyfish, and sergeant majors, yellow-and-black-striped coral reef fish, also subsist on larvaceans.

These voyagers of inner space often carry hundreds of passengers. Many small animals, including flatworms and krill, have been observed resting on larvacean houses and feeding on particles trapped on the webs. Explains Debbie Steinberg, a biologist at the University of California at Santa Cruz who works with Mary Silver, "Larvaceans are like floating food islands; they sustain all kinds of small animals that have hooked appendages for grazing."

Larvaceans are responsible for yet another important connection in the marine food chain. As they spin new dwellings, and when they die, their cast-off houses collapse and sink—sometimes 3,000 feet per day—adding to the flow of natural debris called marine snow. "The sea floor is sometimes littered with clumps of these houses," says Silver. She and Alldredge have learned that the abandoned houses sustain many of the animals living in the ocean's deepest areas. Scientist Bruce Robison, of the Monterey Bay Aquarium Research Institute, refers to the sunken houses as "rocket ships of food."

A closer look at abandoned larvacean houses might help scientists who study Earth's carbon cycle. So

far, they can't find the excess carbon dioxide their calculations say humans have released into the atmosphere by burning fossil fuels. Some process seems to be offsetting the build-up. They do know that phytoplankton turn carbon dioxide into organic matter. If this matter is not consumed at the surface, it eventually falls to the seafloor. "Abandoned larvacean houses may carry large amounts of this organic matter down with them," says Silver. "These houses may be playing a significant role in helping keep carbon dioxide under control."

When they come in contact with another object, *Oikopleura* houses may glow for up to four hours, even after they've been abandoned by their occupants. Scientists once believed this bioluminescence came from microscopic organisms such as dinoflagellates on the house surface. But experiments have shown that houses shed by *Oikopleura* in seawater free of light-emitting microorganisms can still produce light. The larvaceans apparently leave light-producing structures and chemicals behind in their discarded webs.

Not all larvacean houses found in the depths are empty; at least one rare species lives there. More than ten times larger than its shallow-water relatives, *Bathochordaeus charon* spins a web that may reach the size of a pumpkin. This giant larvacean was discovered in the southeast Atlantic by German oceanographers using plankton nets during the 1898 *Valdivia* expedition. Between 1898 and 1991, however, only 17 *B. charon* specimens were found. In the mid-to-late-1960s, Eric Barham of the Southwest Fisheries Center in La Jolla, California, glimpsed several larvaceans he believed were *B. charon* during manned submersible dives off the southern California coast. "Giant larvacean houses were so concentrated in some areas that they were virtually in contact with one another," says Barham, "and when we went deeper in the sub and the vehicle's lights were turned off, numerous objects the size and shape of these houses glowed with bioluminescence."

Barham reported the houses to be in various stages of disintegration, with small crustaceans resting on and swimming around these webs. "But these observations were only possible when the submersible was motionless and the house was directly in front of the viewing port." He suggested that further study of these larvaceans was needed.

In 1993, with the help of Ventana, marine biologists observed several *B. charon* in the depths of Monterey Bay. Rather than forming a house that encircles its body as in other larvaceans, this species spins a fan-shaped, diaphanous veil with fluted edges that floats above it like a huge balloon. According to Robison, "We calculated that *Bathochordaeus charon* can maintain its house at a specific depth for as long as thirty days, collecting particles as it drifts through the water. Cast-off webs of this larvacean that have sunk to the seafloor probably account for a considerable portion of new sediment in the deepest regions of the sea."

In *Twenty Thousand Leagues under the Sea*, Jules Verne described "creatures with umbrellas of opal or rose-pink, touched with a tint of blue; fiery pelagiae, which light our path through the deeps with their phosphorescence." No creatures of his imagination could be more otherworldly than larvaceans whirling through the ocean twilight.

THOUGHT EVOKERS

- What organisms are considered part of the group of animals known as tunicates?

- Explain what Debbie Steinberg means when she describes larvaceans as "floating food islands."

- Discuss the manner by which larvaceans obtain their nutrition.

V ‖ *Plant Structure and Function*

We are, like all animals, linked directly to members of the plant kingdom for survival. Because biotic diversity is the key to maintaining a viable biosphere, continued destruction of these oxygen/glucose-synthesizing systems can only result in negative effects on the earth's faunal communities. But it's not only animals that suffer; plants are affected as well, and such information reinforces the need to maintain diversity among all living systems. One promising way this diversity can be enhanced and maintained is through the science of genetic engineering. Molecular techniques are now available that permit insertion of genes from one plant species into another's genome. However, in our rush to increase plant diversity in this manner, serious environmental consequences might occur. Some microbiologists suggest that plants with artificially introduced traits, such as resistance to pesticides or greater tolerance to natural pests, might produce devastating effects on agricultural ecosystems. The potential benefits are many—increased yields and greater protein production, for example—but scientific and social analyses must be undertaken before any large-scale implementation occurs.

An intact forest contains more floral and faunal diversity than any landed ecosystem, with perhaps the exception of some tropical rain forests. In North Carolina, after years of neglect and devastation, the National Biological Service is funding research to aid in recovery plans for the longleaf pine community. At one time the longleaf forest ranged from Virginia through Florida and along the Gulf Coast to Texas. More recently this original distribution of the longleaf pine ecosystem has been reduced to an estimated 3 percent of its original 60 to 90 million acres. During the 1920s, massive logging practices, turpentine extraction from standing timber, and replacement of the longleaf with pulpwood-producing pines resulted in devastating ecosystem effects. Furthermore, fire suppression activities by federal and state agencies contributed to a marked reduction in the diversity and fitness of the original system. Now, because of increasing awareness of ecosystem dynamics and the economic reality of the longleaf's value, much is being done to enhance and rebuild a national treasure.

14

Longleaf Pine: A Southern Revival

By Tom Horton

Longleaf pine, also known as Georgia, long-straw, yellow, or heart pine, can grow as tall as 100 feet. It can take 50 to 70 years to produce seeds and then can live for 3 or 4 centuries. So re-creating a forest of longleaf—the kind that dominated the coastal plain of the southeastern United States at the time of European settlement—is more than a lifetime's pursuit. But a longleaf revival is under way. The National Biological Service is funding state programs to identify and register the remaining longleaf forests. North Carolina hopes to plant 10,000 new acres of longleaf a year by the year 2000. The Nature Conservancy is making longleaf-forest recovery one of its major initiatives. And the Department of Defense is trying to restore longleaf forests on its southern military bases: 320,000 acres at Eglin Air Force Base, in Florida, and 96,000 acres at Fort Bragg, North Carolina.

Much of this restoration effort is inspired by economics: Longleaf makes a prized straight-growing hardwood. Early country homes were built of longleaf for its unmatched resistance to rot and termites. Among the most resinous of pines (longleaf-pine tar put the *tar* in *Tarheel*), the tree produced a resin that could also be rendered into turpentine—which it was in the late 1800s, by the tens of thousands of barrels. These days pine straw, the thick duff of needles that pads the floor of a pine forest, has become a valuable commodity, harvested for garden mulch in what has become a $150-million-a-year industry.

The greater marvel, however, is the ecosystem of which longleaf pine is merely the most visible component—a community only now gaining wider appreciation, as ecologists probe its workings. An intact forest harbors more diverse assemblages of plant, insect, and animal communities than any outside the richest of tropical rain forests. In a single square meter of one North Carolina longleaf preserve, more than 100 species have been recorded. Deer, black bear, quail, and wild turkey inhabit the woods, which are also home to rare orchids and insectivorous plants. The future of the endangered red-cockaded woodpecker too is bound up with that of this forest, with which it evolved.

The forest once ran in a nearly unbroken swath 200 miles wide, from Virginia down the Florida peninsula and along the Gulf Coast to Texas. The tall trees were widely spaced, allowing light to penetrate the high canopy and settlers to ride horses or wagons through the uncluttered understory of rippling wire grass. Now only tatters of this rich ecosystem remain—an estimated 3 percent of the original 60 to 90 million acres. Bled of their resin

---

Fire is to the longleaf forest what rainfall is to the rainforest. Preservation of the longleaf depends upon learning to replicate the effects of lightning.

---

by the turpentine industry, the trees became susceptible to disease and fire. In the 1920s massive logging followed the westward route of the steam railroads. In the 1930s foresters intent upon raising pulpwood began replacing longleaf pine with plantations of faster-growing species. State and federal governments began suppressing forest fires, thereby short-circuiting the process by which the incredibly fire-resistant longleaf had outcompeted other species for millennia.

What's left is disappearing at a rate of about 100,000 acres a year. Only the merest scraps, no more than 10,000 acres, remain in anything close to the old-growth conditions that an early colonist would recognize.

One of those forests is in Thomasville, Georgia, on a private 200-acre preserve known as the Wade Tract. This may be the closest one can come in the Southeast to tramping through a pre-Columbian landscape, as Thomas Caldwell Croker described it in his 1987 history of the longleaf forest: "Open and parklike . . . the massive trees dotted the rolling coastal plains in a sea of grass. Gentle breezes, laden with a resinous perfume, rippled the longleaf crowns and generated music, soothing to the ear . . . the sweetest this side of the Mason-Dixon line." The Wade Tract is a sort of forester's Rosetta stone, a rare glimpse of the past that researchers can use as a guide in bringing back the longleaf forest.

The Wade Tract is the polar opposite of "deep and dark" often associated with the dense-canopied old growth of other forest ecosystems. Even early in the morning, sunlight floods in through the widely spaced, almost limbless columns of longleaf pines, some approaching 300 years old. One can see for hundreds of yards through the forest and across the rolling, savannalike understory of rippling wire grass.

Reprinted by permission of Tom Horton.

Sharon Hermann, a researcher with Tall Timbers, the private, nonprofit organization that manages the Wade Tract, says that coming here from the midwestern prairies where she used to work was an easy transition, "because longleaf is really a prairie with trees scattered around it."

The forest of the Wade Tract is chockablock with red-cockaded woodpeckers, whose numbers nationwide are down to about 4,600 family groups, from an estimated 400,000 in precolonial times. The birds will nest only in cavities of pines 70 years or older. Some of the nests are so high that they present real climbing difficulties for the biologists who check them for young. Whitish flows of resin, stimulated by the birds' pecking around nest cavities, streak the trunks and discourage intrusion by snakes and other predators.

The key to the recovery of the longleaf ecosystem, says Hermann, lies in the understory. You need not be a botanist to see there is extraordinary diversity here, with hundreds of different plants, from waist-high wire grass to tiny wildflowers, spread thickly in a crazy quilt of textures and shades of green, dotted and streaked with golds, rusts, and pinks.

"We know how to replant longleaf pine, but the real trick is the rest of the plant community. We are just scratching the surface on how to restore it—we don't even know how to get ninety-nine percent of the seeds from plants in the native understory to germinate and establish themselves," Hermann says.

This understory is also quite literally the fuel on which the longleaf ecosystem runs, the basis of what Hermann calls "this wonderfully pyrogenic landscape." Few other tree species thrive on fire like longleaf pine. Frequent forest blazes enable it to maintain its dominance over trees that otherwise would quickly outcompete it.

At some point, says Hermann, she began to think of fire as "not a disturbance here at all . . . more like the rainfall that shapes the rainforest—the defining characteristic of the system.

The absence of fire—that would be a disturbance."

Cecil Frost, a researcher with the North Carolina Department of Agriculture, has been reconstructing the fire regimes that supported the original longleaf forest across the Southeast. By examining historical records and the growth rings of trees, he has come to believe that the bulk of the forest burned every five to six years.

Though Native Americans may have set some of the fires, most were started by lightning drawn to the tallest pines. Dead trees, high in resin and extremely flammable, acted like long-burning wicks and ignited the forest. Once kindled, the understory, especially the dominant wire grass, carried the blaze through the forest. It is likely that only a tiny fraction of strikes actually set fires, but those few fires, in a landscape unbroken by roads, power lines, and a host of other human development, burned for hundreds of square miles at a time.

Not surprisingly, ecologists are discovering that almost every component of the longleaf ecosystem is adapted to—and often dependent upon—being burned. The plants are mostly perennials that invest much of their energy in putting down deep root systems that can send out new growth and seeds after a fire. Wire grass may extend its roots as deep as three feet below ground.

Longleaf seedlings too put down a long, robust taproot. In the so-called grass stage, the little pine tufts appear not to be growing at all. But these delicate-looking clumps are extremely fire resistant. For one thing, the plant in its grass stage grows below the main heat of a fire. The energy stored during this ground-hugging stage enables the young longleaf to boost its sensitive growing tip quickly above the fire-damage zone, growing to a height of several feet in a couple of years and forming a thick, corky, fire-resistant stem.

All the intricacies of fire and the longleaf forest, says Hermann, are only beginning to unfold. Until the last decade or so, even forest man-

agers progressive enough to prescribe regular burning did not recognize how important it was to mimic the natural early-summer timing of lightning strikes. Everything from the flowering and seeding of understory plants to the nesting cycles of forest-floor birds is attuned to fires that occur during the growing season.

So Hermann burns half the Wade Tract every year; she thinks it could be burned for a limited time without harm as often as twice a year. By mid-September the lush understory already almost obscures the charring around the base of the pines, the only clue that Hermann set it all afire as recently as June. Letting lightning do the work is no longer much of an option. "We'll never again have areas big enough to just let nature restore the forest," she says.

But nature has found a perhaps unlikely ally in the U.S. military.

"It looks ugly as hell, but we are trying to make up for lost time," says

---

Commercial foresters often think this forest is sick and dying. But how could they know? Most have never seen what the natural system was really like.

Carl Petrick, a civilian wildlife and forestry manager at Eglin Air Force Base, in northwestern Florida. Petrick and partner Steve Seiber are burning fiercely, as much as 60,000 acres of longleaf a year on the sprawling, half-million-acre base. When they want something burned they bring in all-terrain vehicles with mounted drip torches, or helicopters that drop flammable fuel pods resembling ping-pong balls.

In places, tall longleaf pines rise among the just-burned wreckage of 20-foot-high turkey oaks. Years of fire suppression or ill-timed burning allowed the oaks to shade out any new pine growth. To restore longleaf where decades of misguided management placed slash or sand pine,

Petrick and Seiber are working with the Nature Conservancy on chopping, burning, and replanting.

Conflagrations are nothing new among the forests of Eglin, where the military develops and tests munitions, optical guidance systems, Smart Bombs, and Hellfire missiles that can melt through a tank as if it were butter. Ironically, fires set by guns and bombs and troop movements have for decades played an inadvertent role in keeping substantial tracts of longleaf in relatively good condition. So Petrick and Seiber command 400,000 acres of forest, some 320,000 of it either in longleaf pine or suitable for its restoration. They call it their ecological battlefield.

This forest, drained by some 830 miles of streams, rivers, and tidal estuaries, harbors some of Florida's largest populations of black bear and red-cockaded woodpecker. Eglin's 724 square miles embrace half of the 83 distinct plant and animal communities that occur in Florida. Several thousand acres of the base, some lying in the "ricochet zones" of munitions-test ranges, are nearly as fine an example of old-growth pine as the Wade Tract, perhaps the largest stands of such quality remaining in the world.

Eglin, Petrick says, has been the first Department of Defense installation to wholeheartedly embrace the concept of ecosystem management that is being touted, with varying results, by

the Clinton administration. But Eglin is not alone. Fort Bragg, near Fayetteville, North Carolina, is managing 96,000 acres of longleaf, which include some of the best pockets of rare plants in the state and a major population of red-cockaded woodpeckers. Other bases, such as Fort Jackson (Columbia, South Carolina) and Fort Benning (Albany, Georgia) are also beginning to manage big acreages of longleaf.

Ironically, military managers sometimes have a freer hand in restoring their forests than does the U.S. Forest Service on its own large holdings across the longleaf's range. Researchers say they continue to be hampered by unrealistic goals for timber sales and multiple-use management.

For example, "most of the federal money for fire management on public lands is for fire suppression," says Hermann. "A lot of the burning they do [for longleaf] is done as fuel reduction" (preemptive burning to reduce the likelihood of a conflagration).

"We're not out in front because we're so fast and smart," says Petrick. "We just don't have our shoelaces tied by mandates that run counter to managing for diversity. What we are doing is managing first and foremost for the ecosystem." Everything else, he says, from timber and game species to the endangered woodpeckers that the military legally must protect, "is a by-

product. We do what is best for the ecosystem, then see what we can do with the by-products." Besides hosting a large population of woodpeckers, the base cuts and sells $750,000 a year in timber and permits public fishing and hunting in places.

For its efforts, Eglin has received awards in the past few years from the Sierra Club, the Nature Conservancy, and the National Wildlife Federation. It also won the Defense Department's top awards for environmental management.

When the Forest Service published Croker's report in 1987, a history seemed all that would soon be left of the species. Now the longleaf's decline is stopping and may be even reversing on public lands across most of its range. On private lands, which contain about two-thirds of all longleaf acreage, the prospects are less clear but possibly improving, as large, commercial landowners like Georgia-Pacific begin to pursue limited restoration.

Longleaf advocates say that even if commercial forestry's renewed interest falls short of ecosystem management, that approach keeps options open a lot better than the prevailing emphasis on pulpwood plantations of loblolly and other quick-growing pines.

The Woodpecker That May Save Its Own Woods

The federal recovery plan for the endangered red-cockaded woodpecker has helped focus attention on the longleaf pine and hastened its restoration on public lands. But on private lands the plans have sometimes hastened cutting of the older trees preferred by the bird. At best, landowners' concerns are holding back any widespread move toward longleaf.

That may change. Ralph Costa, the U.S. Fish and Wildlife Service's woodpecker-recovery coordinator, says he is rewriting the plan to allay the bulk of landowners' concerns. His ideas range from compensation for not being able to cut timber around nest sites to tax breaks and even "woodpecker credits" that would allow timbering once a site had contributed so many young woodpeckers for transplantation to other areas.

The private Environmental Defense Fund is working along similar lines to defuse the current property-rights backlash toward

woodpeckers and other species. "We are seeing more actions to prevent endangered species' occurring on property," says Melinda Taylor, director of the fund's North Carolina office, in Raleigh.

In the strictest sense, the federal recovery plan does not ensure that the longleaf ecosystem will remain intact, since the woodpecker is known to need only old trees, not the diverse understory. (The old-growth Wade Tract doesn't even meet minimum federal standards for the red-cockaded woodpecker, which are based on the size and number of all the tree trunks per acre. Yet the bird flourishes.)

Still, Costa and some ecologists suspect that an intact ecosystem may be found to be most beneficial in the long run. "Any way you look at it, the woodpecker is going to drive this [longleaf] recovery a lot more than it's going to retard it," Costa maintains.

Tom Horton

Longleaf "is a more difficult animal to work with than other pines," concedes Bob Weir of North Carolina State University's Department of Forestry. A loblolly seed orchard produces 500,000 seedlings per acre per year. Longleaf averages only 75,000. It produces seed in large quantities only every three to seven years—a still mysterious timing that, when it does happen, occurs simultaneously across most of the species' range.

Longleaf is also more difficult to transplant because of its long taproot, and seedlings cost up to three times as much as those of loblolly. "But remember," Weir says, "we've invested perhaps forty-five years and fifty to sixty million dollars in genetics and growing of lobs, and less than a decade and two hundred thousand dollars in longleaf."

Management for quail, deer, and turkey hunting, combined with selective timbering, can produce a forest-scape and a fire regime very similar to that of the Wade Tract. In fact, a couple of thousand acres thus managed surround and buffer the Wade Tract. North Carolina's 50,000-acre Holly Shelter Game Lands are another fine example of such management.

It is hard to know how fast the lessons from Eglin, the Wade Tract, and other research sites will translate to restoring the longleaf forest. Sharon Hermann says foresters still have a great deal to learn. "Commercial foresters I bring here go into spasm. 'Your forest is sick, dying, understocked,' they tell me—but how could they know? Most of us have never seen what the natural system used to be like."

THOUGHT EVOKERS

- How was the longleaf pine ecosystem fundamentally destroyed?

- Why is fire necessary for longleaf pine survival? Explain.

- Describe the steps being taken to rebuild the longleaf pine ecosystem.

15

A Tree Too Tough to Kill

By Bil Gilbert

In some ways the mesquite tree is the botanical counterpart of the coyote. Like that mammal, the mesquite is not particularly grand when compared with others of its order—oaks, cottonwoods, madrones. However, both mesquites and coyotes are remarkably tenacious and successful. They have enormous ecological impact on everything around them.

The two species also have a strikingly similar relationship to mankind. For most of this century, over millions of acres of arid southwestern lands, we have been expending considerable amounts of money, time, and ingenuity trying to eradicate them or reduce their numbers. Ranchers and the state and federal public servants who represent ranchers' interests have seen—and generally still see—the two as ferocious enemies: the coyote because it may prey on livestock and mesquite because it may "prey" on grasslands that feed livestock. There is serious disagreement about whether the massive and continuing campaign against the two is in the general economic and environmental interest. However, there is little dispute that in this competition we have not yet achieved much more than a shaky draw. Despite our activities—or because of them—mesquite has done very well in areas where we first found it and has steadily pushed its way into new regions.

Paradoxically, because mesquites and coyotes have a well-proven ability to survive, we have come to admire them, grudgingly, as worthy opponents. This is particularly true in regions where they are most prevalent and where people have tried long and hard to rub them out. Mesquites and coyotes have become the principal living symbols for what is left of the Wild West and a source of sour pride for the inmates of the region.

> **Mesquites and coyotes have become the principal living symbols for what is left of the Wild West and a source of sour pride for the inmates of the region.**

Whether you consider it good or bad, mesquite is a tree of remarkable properties. It is, in taxonomical terms, a Leguminosae—of the pea family, which includes the locust, acacia, mimosa. Paleobotanists believe mesquite originated in South America perhaps fifty million years ago. Since then it has been on the move both structurally and geographically, and there are now some fifty species. Four are considered natives of the Mediterranean littoral, probably having arrived as marine flotsam. Other mesquites in Africa, Asia, and southern Europe were brought there by humans.

In the United States there are two principal species: honey mesquite (*Prosopis glandulosa*), found from the Texas gulf coast to New Mexico, and velvet mesquite (*P. velutina*), whose range extends across Arizona into California. The northern limit of the tree's range runs along the northern boundary of Oklahoma, extending westward through southern Colorado, Utah, and central Nevada. North of that line, there is too much frost for mesquite. For the same reason, mesquite is seldom found to the south above 5,500 feet and does best below 4,500.

Otherwise it is remarkably adaptable, growing on flats and precipitous slopes, in sand, gravel, rocks, loam, and clay, and in alkaline, saline, acid, and neutral soils. It can tolerate full desert heat and survive in areas where there is as little as six inches of rainfall a year. Yet it will also grow in places where there are one hundred to one hundred fifty days of frost a year and thirty inches of precipitation. Mesquites show a great variety of growth habits. On hummocks in sand-dune country they will sprawl along the ground. In other places they stand in bushy thickets. Under the most favorable conditions mesquites become proper trees, forty to sixty feet tall with a trunk diameter of three feet or so.

Mesquite flowers in early spring, and even if there is a late cold snap and portions of the tree are frost-killed, dormant buds below the dead-wood are activated and shortly send up replacement shoots. Occasionally, these dormant buds may sprout even if the tree has not been injured, and mesquite will sometimes flower two or three times during a summer. The elongated fruit pods resemble those of the black-eyed pea, and each bears fifty to sixty seeds. Each seed is encased in a hard, epoxy-like covering that permits it to remain dormant but viable for years.

The mesquite's roots are extensive and contribute greatly to its success. A lateral root system, a few feet below the surface, fans out from the trunk of a mature tree as far as sixty-five feet, and its formidable taproot may penetrate downward for an equal distance.

There is a common belief that these taproots reach the water table and continually pump from it—to the detriment of the flora and fauna closer to the surface. This perception of hydrological greed accounts for much of the hostility toward mesquite. In fact, mesquite roots seldom reach the water table, and when they do they draw from it only in times of severe drought. Apparently the taproot serves as a kind of emergency back-up system, enabling the tree to survive in arid places and conditions. Even during a drought, the mesquite taproot has little effect on water available to small grassland plants, since it is using water these shallow-rooted species cannot reach anyway.

When water is plentiful, mesquites seem to be "luxury" water users, sucking up a lot of it through their lateral roots, transpiring rapidly, and growing at an impressive rate. In controlled experiments conducted in Arizona, where water was easily available, mesquite used 1,725 kilograms to produce one kilogram of dried plant matter. To make the same amount, grama grasses used only four hundred to five hundred kilos of water. However, under drought conditions, grasses and other forage plants continue to transpire at the same rate, while mesquite reduces its transpiration rate and water use. Growth declines or halts, and the trees enter a resting mode, which they can maintain without permanent ill effects for long periods.

Mesquite is often thought of as a desert species, but it is not a true xerophyte, like cactus, because it apparently evolved in the wetter grassland. However, it is able to endure in very dry areas or in periods of drought as few other grassland species can. Thus, in regions undergoing desertification, mesquite will take over.

The Sonoran Desert between Tucson and the Mexican border illustrates both mesquite's natural tendencies and its adaptability. On the flats and gravelly ridges mesquite is plentiful, generally the most sizable woody plant even though it is a fairly scraggly bush. Its stems are spindly, and there is a yellowish, sere look to its summer foliage. Barrel cactus, cholla, and ocotillo are common associates. However, every few miles the desert is cut by winding ribbons of bright-green foliage marking the bottoms of dry stream beds and erosion washes, where there is more moisture closer to the surface. Here, mesquites grow as substantial trees. Also, mesquite is a disturbed-land species: Its seedlings and young trees get a particularly good start on ground that has been broken up and stripped bare of competing species. In grasslands, erosion caused by torrential summer rains and flash floods is the most common natural creator of disturbed lands.

Like its relative the eastern locust, mesquite not only pioneers in bare places but tends to improve them. Mesquite roots, like those of many Leguminosae, transfer nitrogen to the soil and stabilize it. The summer foliage provides shade and slows down evaporation, and when the leaves fall in October they add organic matter to the soil. For a tree like the locust these restorative acts are in a sense suicidal. As the disturbed land is restored, other species—maples, oaks, tulip poplars—establish themselves and eventually crowd out the locust. Mesquite, on the other hand, is often both a pioneer and a dominant

forest species, since there is enough moisture in the bottom of the erosion washes to satisfy its requirements but not enough to encourage larger trees. In southern Arizona, cottonwood, walnut, and sycamore share the same general range with mesquite but are usually found only where water is very close to the surface.

Archeological and anthropological records indicate that mesquite was of vital importance to the early native residents. The trees provided shade, firewood, and building material, and the fibrous bark was woven. Mesquite pods and seeds—whose protein content is higher than that of soybeans—were eaten raw, roasted, dried, ground into flour, and used to brew teas and fermented drinks.

The pods themselves taste sweet, which makes the fruits particularly attractive to large foraging animals. Therefore mesquite seeds—which usually are impervious to digestive processes—are carried in the stomachs of browsing mammals and eventually scattered in their dung. (A researcher found that a single cow chip may contain as many as a thousand viable mesquite seeds.) Creatures who feed significantly on mesquite include cattle, horses, sheep, goats, deer, javelina, antelope, all manner of rodents, coyotes, raccoons, coatimundis, and skunks, as well as quail, turkeys, and ravens.

Judging by the casual observations of Anglo-Americans who first reached the Southwest in the forepart of the nineteenth century, mesquite was found in approximately the same regions as it is currently, but there was a major difference in how it was distributed. Then as now there were stands—bosques—of large mesquites in the draws and washes. However, in the great seas of grass that stretched across Texas, New Mexico, and southern Arizona, there were few mesquites, generally stately specimens thriving in a benign, accidentally created micro-environment. They provided welcome shade and served as prominent landmarks in the savannahs. Today, relatively low, scrubby mesquite covers ("infests" is the word often preferred by range managers) some seventy million acres of the Southwest, having replaced much of the former grassland.

There is little dispute that this change was caused by the coming of the Anglo-Americans—particularly the cattlemen. Longhorns were moved into the savannahs by the hundreds of thousands. Unlike the native buffalo, which were migratory, the cattle were held in one area year-round. In grazing they stripped away the forage plants, and their hooves churned up the thin topsoil. When one area was laid bare, the cattle were driven farther west. And as they traveled they distributed the mesquite seeds they had ingested. It is still possible to identify old cattle trails because of the dense mesquite thickets along them.

Another factor that contributed to the spread of mesquite: Trees up to five years old are vulnerable to fire, which regularly swept through the original prairies. Once mature, though, mesquite is very fire resistant. Tops may burn, but dormant buds survive at ground level to send up new shoots. To protect property, stock, and forage, ranchers attempted to prevent and fight wildfires. To the extent they were successful mesquite benefited. More important, overgrazing reduced the amount of inflammable litter and thus made fires less frequent and extensive. Once the presence of cattle let mesquite gain a foothold in the original savannahs, the remaining grasses were at a disadvantage.

In the 1880s and 1890s some five million cattle were driven west out of Texas into the virgin prairies of southeastern Arizona. Shortly after their arrival the area suffered a severe, prolonged drought. Desperate cattle, thousands of which died each year of thirst and famine, had by the early twentieth century devastated much of the range between the Chiricahua Mountains and the Gila River. The land became cross-hatched with new erosion washes and rapidly proliferat-

ing mesquite thickets. Only in the past quarter of a century has better range management begun to repair some of the damage from this period.

The remaining cattle were moved and in the 1920s came to the Altar Valley, which lies west of Tucson under the eaves of the Baboquivari Mountains. Since desertification and the spread of mesquite occurred there fairly recently—and is still occurring—the Altar makes a good place to observe these processes. When I visited in 1983, it was also serving as an extensive field laboratory for Linda Leigh and Dennis Cornejo, two University of Arizona graduate students who were among the region's most knowledgeable scientists in these matters.

Leigh was dreaming of getting access to a backhoe to be better able to examine mesquite root systems. Her academic work involved investigating the growth rates of the trees; she also worked with the Office of Arid Lands Studies at the university. Cornejo, a broad-spectrum desert ecologist, was, at the time, research associate at the Arizona-Sonora Desert Museum. He proved to be a generalist, given to freewheeling speculation and unexpected modes of expression. "What happened," said Cornejo of the spread of mesquite during the last century, "was that the trees were just lurking in the washes waiting to pounce. They were biding their time until something came along and disturbed the land to let them get up in the flats. Cattle did that, disturbed the land and let them out."

"Dennis," Leigh said reproachfully, "remember when you say things like that they may end up in print."

In the middle of the Altar Valley lies a broad, shallow wash that was there long before the coming of the cattle. According to early records, the wash often flowed during the summer, when the area gets most of its precipitation—about eleven inches annually. Often it spilled over the banks, creating a wet, almost swampy area that might persist for several

months. Early homesteaders recall that the grass stood hip high in the valley and—except along the washes themselves—there were only a few big mesquites standing in the flats.

Conducting a before-and-after tour of the areas, Leigh and Cornejo led the way up a tributary wash to a place where some years ago it had been filled with an earthen dike to catch runoff from the Baboquivaris and create a shallow, intermittent stock tank.

It was late spring, and the air temperature was fairly moderate for these parts, 100 degrees Fahrenheit. There was a foot or so of water in the pond. Around it was a fine mesquite bosque, the trees standing thirty or forty feet tall. A breeze was blowing through their lacy foliage. Spreading out from the wash, in the flats, was a lot of mesquite scrub, some growing in two- or three-acre thickets so dense that a man on foot could hardly have penetrated them. Javelina had less trouble—one flushed from such a tangle skipped agilely through it, snorting as it went.

The trees on the flats were low and bushy. "There are essentially two kinds of mesquite," Cornejo said. "Back there," he nodded in the direction of the forest above the pond, "you have your happy mesquite. Here," he indicated the thicket, "is unhappy mesquite." Leigh was shocked by this taxonomical frivolity.

Beyond the thickets the low mesquite tree-bushes stood alone or in scattered clumps. Between them the predominant growth was skunk bush and burro weed, two low, leathery, arid-land species almost as tenacious as mesquite. But unlike it, they provide little forage for animals. Interestingly, what grasses remained were concentrated in small oasis-like communities under the mesquites, whose roots were holding some topsoil in place and whose leaves were providing shade and mulch. In this particular location mesquite was not the enemy but the last friend of the grasslands.

Finally there were "balds," patches of an acre or two in which the ground was bare and almost as hard as macadam. A few dying or dead mesquites were there, but very little else. The trees moved in when the ground was first disturbed. The grasses and other low-growing plants were so quickly uprooted that not even the mesquite could stabilize the remaining topsoil. Summer rains swept it into the Altar Wash, leaving only the caliche hardpan, which soon baked to an adobe consistency. The rate of runoff and erosion accelerated. Unless something reverses the process these balds soon will be sterile deserts.

As is true elsewhere in Arizona, the ranches in the Altar Valley are made up of private and public lands leased from state or federal agencies, in this case the Bureau of Land Management. Both private and public land managers are well aware of what is happening and have taken at least some corrective steps. Overgrazing is no longer a problem, and in the study area used by Leigh and Cornejo there have been no cattle for the past five years. However, there are doubts about whether this land can restore itself soon, even if left alone. Leigh pointed out that there may not be viable dormant seeds left in the cracks of the hardpan. She wanted to have earth samples analyzed for seed content, but resources for this work have not been available.

In the southwestern ranch country there are two main complaints against mesquite: first, that cattle cannot graze in the heavy stands, and second, that mesquite ruins pastures by shading out and stealing water from grasslands. The second complaint is difficult to evaluate. Mesquite thickets can dominate and displace low-growing vegetation. The plant obviously uses water. However, eradicating mesquites in non-thicket areas does not automatically make the water they have been using available to forage plants. The effect depends upon a number of other conditions. In some cases moderate stands of mesquite—fifty or fewer trees an acre—may have a positive influence on grasslands, improving their moisture-holding capacity.

The reaction of ranchers to mesquite usually depends upon how much of it they have. C. R. Barbier has worked with cattle all his adult life and presently manages a fifty-thousand-acre ranch in the high Sonoran grasslands of Arizona a few miles north of the Mexican border. "I have worked in places where the only way I wanted to meet mesquite was from the seat of a bulldozer," Barbier says. "But up here I wish we had more mesquite. Besides being pretty and good for wildlife, it makes nice shade. Another thing, the dry season comes in the spring and early summer here, when there isn't much green or growing but mesquite. In some places I know there are two or three bad weeks when mesquite is about all the cows have to eat. It just depends, whether mesquite is good or bad."

There are, however, extremists on the subject, ranchers who can work themselves into a rage when it is suggested that, if brush is a problem, it is a problem created by cattlemen. Their view is that the only good mesquite is a dead one, that it should be eradicated completely, and that when this is done on leased public lands, which most ranchers use, the public should pay for the job. This "nuke the mesquite" approach, as Dennis Cornejo calls it, has some strong supporters in conservative circles. But even if such an ultimate solution were desirable, there is no reason to believe we can eradicate the trees, short of sterilizing seventy million acres or so of the Southwest. At a practical level, ranchers and public agencies cope with mesquite in one of two ways. One method is to keep the brush thinned down and out of unoccupied areas. The other is to consume it—to find new and profitable uses for mesquite.

Cowboys, using the first method, have been hacking away at mesquite with axes, saws, and brush hooks for generations. In recent times these hand tools have been supplemented with earth-moving and gouging machines. But even with them, grubbing out mesquite remains a hard, time-consuming, and fairly costly operation, and dormant buds will send up new growth if even a few inches of stump remain. The same thing will happen if mature mesquites are burned. So each tree or bush must be uprooted.

. . . grubbing out mesquite remains a hard, time-consuming and fairly costly operation, and dormant buds will send up new growth if even a few inches of stump remain. The same thing will happen if mature mesquites are burned. So each tree or bush must be uprooted.

An early and still effective eradication method is to cut the trees down and douse the stumps with diesel fuel to kill dormant buds. This works and does little damage to the surrounding floral community but is labor intensive and fairly expensive. Since World War II, herbicides have been tried for mesquite control. The consensus is that the best, in terms of effectiveness and cost, is 2,4,5-T, a phenoxy, dioxin-containing compound that defoliates a variety of plants by altering their metabolic balance. It was one of the ingredients in Agent Orange, which became infamous during the Vietnam War. The association so alarmed the public that the Environmental Protection Agency withdrew the herbicide from general agricultural use. There is currently considerable interest in using tebuthiuron as a replacement. It does not contain dioxin and kills by disrupting photosynthesis. According to Eli Lilly—which markets tebuthiuron under the trade name Grasslan—and outside experimenters, it has no direct effect on fauna when properly used.

Howard Morton, a senior scientist at a research station maintained by the U. S. Department of Agriculture in Tucson, is a leading authority on herbicidal control of brush. He previously worked with and thought well of 2,4,5-T and feels that tebuthiuron may be the best available substitute in certain situations. It is customarily delivered from the air via small pellets in quantities that can kill a wide range of plants. However, Morton says the number of pellets dropped can be controlled very precisely and at a rate of about one pound per acre, tebuthiuron gets mesquite but leaves most of the grasses. He has been testing the compound on experimental plots in this country and Mexico for most of the past decade and reports that it kills at least eighty percent of the mesquite. He says it works better in thin, shallow soils than in heavy, deep ones and that most of the shallow-soil study plots he treated eight to ten years ago are still largely free of mesquite. As to residual effects, Morton recommends that no new seeding—generally done after brush is controlled—be undertaken for eighteen to twenty-four months after tebuthiuron treatment.

The Coronado National Forest occupies almost two million acres or so of southern Arizona and includes sizable tracts of grazing land leased to local ranchers. In the summer of 1982 Coronado officials announced a plan to use tebuthiuron to control mesquite on a thousand acres of forestland, part of a twenty-thousand-acre leased tract near Arivaca, a small rural community at the head of the Altar Wash, some fifteen miles north of the Mexican border. There were immediate objections. Officials from the Arizona Game and Fish Department expressed concern about what would happen to fairy duster, a mesquite relative that provides especially good wildlife forage. The Forest Service agreed not to spread tebuthiuron on three hundred acres originally scheduled for treatment, but this did not halt the argument. Arivaca area residents organized protest meetings, wrote a lot of anti-tebuthiuron letters to Tucson papers, and protested directly to the Forest Service. Eventually two hundred residents—virtually all the non-ranchers in the area—signed a petition

asking that the tebuthiuron plan be scrapped.

A leader in this campaign was Barbara Stockwell Nachbaur, a commercial beekeeper living near Arivaca. She says that although she has an obvious business interest—mesquite and fairy duster provide most of her bee pasture—she has strong feelings against most herbicide programs: "We are too dependent on chemicals in our lives. We look to them for easy solutions to immediate problems and then later find out we have created new and worse problems because we used them. If mesquite has to be controlled, I am more in favor of mechanical means. A bulldozer causes damage, but the vegetation comes back quickly, and at least there is a human brain running the machine. A lot can go wrong with spraying from the sky. Nobody knows what is happening on the ground."

Larry Allen, a Forest Service range manager, was to be in charge of the Arivaca herbicide project. He feels tebuthiuron is a good, effective, safe compound and that objections to it are more emotional than factual. Last winter he remarked publicly that he saw "no biological or environmental reason not to apply tebuthiuron" in the Arivaca area but added, "We do see a real public relations problem."

Largely because of this, the Forest Service has not used tebuthiuron near Arivaca and now says the plan is undergoing intensive review. Nachbaur and the other protesters remain distrustful and are paying close attention to what the service is doing. They believe chemical control is unnecessary because there is now a better method of dealing with mesquite: the land imprinting technique developed by Robert Dixon.

Like Howard Morton, Dixon is a scientist at the Department of Agriculture's Tucson research center, but he works more as Morton's opponent than his colleague. Dixon received his early academic training at the University of Wisconsin, where he was influenced by Aldo Leopold's land ethic. He is suspicious about the consequences of agricultural depen-

dence on herbicides. His special professional concern is over the way we have been digging up the ground, disturbing topsoil (the loss of which he considers to be one of the worst troubles facing mankind), and making the land more vulnerable to erosion and often less fertile. This is particularly the case in the arid lands with which Dixon has been involved during the past twenty years.

Dixon has invented an alternative to digging up the land that he believes could make dry-land farming more environmentally benign. He calls it a land imprinter. It is a heavy roller, like those used in highway construction, dragged by a caterpillar tractor. The roller is set with big pyramid-shaped teeth, four to six inches long, that are welded to it in an irregular pattern. The positioning is crucial, since Dixon feels that any device that makes straight-line cuts in the ground creates potential watercourses and therefore erosion problems.

A pass with the roller leaves the earth looking like a huge, free-form waffle. The teeth do not overturn the topsoil but press it down into the holes, which are wide at the top and taper to a point. These, says Dixon, serve as miniature catch basins, gathering and holding water in place. Furthermore, as the tractor and roller proceed they chew up and crush vegetation and pack it into the holes, where it becomes moisture-retaining mulch and eventually organic plant food. This not only improves the fertility of the land but creates areas where seeds, either in the soil or sown simultaneously, will germinate quickly.

For mesquite control, Dixon feels that imprinting combined with good management usually is as effective as and more environmentally prudent than herbicides. Nurtured in the funnel-like impressions, new growth is rapid, and given this kind of head start, forage plants crowd out mesquite shoots, as they did when the prairies were virgin. If properly grazed, such growth will in time create litter that will burn, and the fire will further keep brush under control. Dixon calls the land imprinter a re-creation of the buffalo, acting on the land as the hooves of the migratory herds did. He feels that with the technique it is possible in given areas to reestablish the savannah environment—seas of grass set with occasional, useful groves of mesquite.

A California firm is manufacturing and marketing commercial land imprinters. As a result of a trip to China, Dixon is also tinkering with models of manually operated imprinting tools for areas where labor is plentiful and cheap.

Dixon claims that where land imprinting has been used—in Utah and Texas, among other places—it has favorably impressed farmers and ranchers. Ralph Wilson, who has a sixty-thousand-acre ranch near Tucson, worked with Dixon to make his own land imprinter and has used it on about four hundred fifty acres. The first test area was imprinted in early spring, and Wilson was able to move cattle into the pasture's knee-high grass by the following fall. He estimates the cost of imprinting and seeding at between twenty and twenty-three dollars an acre. Mesquite control with tebuthiuron, and subsequent seeding, costs between twenty-five and thirty-five dollars an acre.

Although Wilson found imprinting excellent for regenerating grasslands, he did not find it particularly effective for controlling heavy brush. Trees and bushes mashed down by the roller re-sprouted. However, Wilson's ranch is only moderately infested and he wants to keep some mesquite: "In bad drought years my cattle forage on the trees during May and June. It is what gets them through. I think one of the worst things a rancher could do is get rid of all his mesquite."

Robert Dixon and his land imprinter have been evoked frequently by environmentalists as an alternative to herbicides, but one that local land managers won't try because they have such a mind-set in favor of chemicals. Dixon, who is not publicity shy, has done nothing to discourage these feel-ings. He and his work have often been cited, favorably, in the Tucson press. In his public comments Dixon, in addition to promoting land imprinting, has been fairly critical about tebuthiuron, herbicides in general, and those who want to use them.

The "nuke the mesquite" ranchers see land imprinting as something that will keep them from killing brush in whatever way they want. Pro-herbicide land managers think Dixon has made what should be a technical dispute a public one, is rocking the administrative boat, is a fanatic about disturbing the land, and is a publicity hound. The technical complaint about land imprinting is that it does not clear the land, that well-established trees will re-sprout much faster than forage plants grow. Also it is claimed that a land imprinting rig cannot mash big trees or be operated in rough country. Larry Allen, the Coronado Forest range manager, says that land imprinting was not considered an alternative to tebuthiuron in the Arivaca area because "it will not work. It won't take out that kind of mesquite."

Dixon counters that his machine will go anywhere a caterpillar tractor can, that big trees should be left in place to create a savannah environment, and that land imprinting is generally a good mesquite control method. He says it is not being used because the Forest Service and his own Department of Agriculture research colleagues are inflexibly pro-herbicide. "I was told," says Dixon, "that what I was doing was interfering with policy." And what is Dixon's understanding of the policy? "Obviously to eradicate mesquite with herbicides."

Because of his promotion of land imprinters, Dixon claims he was suspended for a few days without pay; that he is under pressure to resign or transfer to another area; that his research funds and staff have been drastically reduced. He thinks the source of most of his troubles is Howard Morton, who, in addition to being the tebuthiuron tester, is Dixon's administrative superior.

Morton is more conciliatory. He says he thinks that in certain areas land imprinting is useful and he would like to see more experimentation with it. He says he does not care to make any comment on Dixon's charges about professional and personal harassment.

Long before tebuthiuron, land imprinters, and southwestern ranching were invented, original residents of the region thought mesquite was desirable and enjoyed using various parts of the tree. Now that a century of destruction has been only minimally successful, there is a movement to return to the old ways. This approach is called "mesquite utilization."

Even when anti-mesquite feeling ran highest there was agreement that it made fine firewood and in many areas was the only firewood. Roasting meat over mesquite coals became another part of the western legend. A number of entrepreneurs have capitalized on this folk image: Bags of mesquite chips, charcoal briquettes, and nuggets are being exported to suburban barbecuers as if mesquite were a kind of woody Coors beer. Since mesquite cutting is still mainly a cottage industry, there is no way to estimate how big this business is or may become. About 2,500 cords a year are now being cut in Arizona, twice as many as ten years ago. I know of a factory in Westminster, Maryland, that is bagging mesquite bits and pushing them in the Baltimore-Washington area, one of the best American habitats for backyard cooks. Also, an acquaintance who lives near Fairbanks, Alaska, sends periodic and desperate requests for mesquite suitable for grill work, claiming it is the only thing that does justice to a nice piece of moose or bear.

In October 1982 Texas Tech University sponsored a Mesquite Utilization Symposium. Many of the uses discussed there have become economically attractive in recent years because scientists at Texas Tech have developed a field model of a mesquite "combine," a machine that rolls through the brush, cuts down trees up to six or seven inches in diameter, and reduces all the parts to wood chips at an operating cost of $7.50 per ton of green matter. Pulverized mesquite can be used in making particle board, as feed and roughage for livestock, and as commercial fuel. In most cases its price is competitive with other materials used for such purposes.

Texas Tech nutritionists suggested that the sweet-tasting flour made of mesquite pods and beans has a high potential for human use, especially when mixed with other cereal grains. Mesquite wood is hard, strong, and dense, and it has an irregular grain pattern that makes it very attractive, but it has seldom been used for commercial woodworking. Two researchers from Northern Arizona University have been milling and drying mesquite lumber. They distribute it to commercial cabinetmakers who, it is reported, would be eager to use it if there were a dependable supply. Though there is a certain prejudice against mature mesquites as weed-varmints, they are attractive trees of graceful shape and foliage. These qualities, plus their adaptability, should make them desirable ornamentals—and practical ones in the Southwest where homeowners can often be found pouring precious water on spindly maples, birches, rhododendrons, and other imports. It was reported at the Texas symposium that a few creative nurserymen and landscape architects are beginning to work with mesquite. Trees are being collected, balled, pruned, and sold for a hundred dollars or so. Increasingly, ranchers with big mesquite thickets that provide excellent wildlife cover are leaving them in place and charging hunters for their use.

Techniques for utilizing mesquite and markets for mesquite products are in the first stages of development. However, there are so many economically promising ideas in this line that consumption in one form or another, rather than eradication, may well be the wave of the future. In time, southwestern ranchers may come to regard a mesquite woodlot much as people do a sugarbush in Vermont or a stand of black walnut in Missouri.

THOUGHT EVOKERS

- Describe the adaptations intrinsic to mesquite that have given it the ability to withstand various attempts at eradication.

- Explain this statement: "In some ways, the mesquite tree is the botanical counterpart of the coyote."

16

Ecological Risks of
Genetic Engineering of Crop Plants

*Scientific and social analyses are critical to realize benefits
of the new techniques*

By Carol A. Hoffman

Molecular biologists working to improve crops often state that the ecological risks from releasing genetically engineered (transgenic) organisms are small and point out that breeders have introduced new genes into organisms by crossing with wild relatives for decades with no apparent harm. Brill (1987) states that biotechnology in fact reduces risk because traits are more tightly controlled in these crosses, that is, single genes are transferred into the crop instead of blocks of genes.

Traditional breeding techniques usually rely on crosses with closely related plants, either subspecies or wild ancestral species of the crop, because incompatibility barriers often prevent crossing between more distantly related gene pools (Harlan and DeWet 1972). Incompatibility presents no barrier with molecular engineering techniques, and, in fact, genes can be introduced from a closely related species, distantly related species, or even an organism from outside the plant kingdom.

It is the novelty of the introduced trait (in terms of the absence of evolutionary history with the plant or interacting species in the environment) that distinguishes crop breeding using molecular genetic engineering techniques from traditional breeding methods.

Until recently, the predominant ecological concern was that plant taxa with novel engineered traits would become pests in the environment. Studies of crop-weed comparisons have shown that plants can evolve invasive genotypes based on a few major gene polymorphisms, which are usually followed by physiological or life-history changes.[1] The probability that this change to weediness will occur with transgenic crops is unknown.

The direct evolution of transgenic weeds, however, is not the only or even primary concern with the release of plants. Crop plants are capable of transferring genes, by hybridization, over relatively long distances to related plants that differ markedly in their life-history characteristics. Many of these crop relatives have the weedy characteristics of high reproductive output and seed dispersal. Thus the genes carefully inserted by the molecular biologist may be further transferred into new organisms by the plants themselves.

This issue of *BioScience* focuses on the environmental risks associated with the development and deployment of genetically engineered crop plants. These articles are based on talks presented at a symposium at the 1988 AIBS annual meeting in Davis, California. They address three topics: the special risks associated with transgenic crops, existing data on patterns of gene transfer between crop plants and their wild relatives, and techniques that can be used to address experimentally questions of risk.

Techniques and Traits under Development

Initially, successful gene transfer using genetic engineering relied exclusively on the use of *Agrobacterium* as a gene vector. The trait of interest was inserted into an *Agrobacterium* plasmid, the plasmid crossed the plant cell boundary, it inserted the

> **Genes carefully inserted by the molecular biologist may be further transferred by the plants themselves.**

genetic material into the DNA of the host plant, and the newly transformed cells were cultured into whole plants (Grierson and Covey 1984). This technique worked successfully with plants of the Solanaceae family, such as tobacco, petunia, tomato, and potato. For this reason, commercial development of crop varieties of these species has proceeded quickly. Many companies (e.g., Monsanto, Agracetus, and Calgene) are already in the field-testing stage with their plants.

Although potatoes and tomatoes are major commercial crops, grain crops, such as rice, wheat, and corn, have a much greater potential for worldwide sales (see table). For monocots and other plants, the *Agrobacterium* vector

Carol A. Hoffman is an assistant research scientist in the Institute of Ecology, University of Georgia, Athens, GA 30602.

---

[1]S. Jain, 1988, presentation at AIBS meeting in Davis, CA. University of California, Davis.

From *BioScience*, vol. 40, no. 6, pp. 434–437. © 1990 American Institute of Biological Sciences. Reprinted by permission.

approach has been unsatisfactory and difficult (but see Raineri et al. 1990) and other genetic engineering techniques are being developed (Strange 1990).

Electroporation is a technique that uses electrical current to open holes in the plant membrane and allow passive transfer of genetic material into the cell. This technique shows promise for several grasses such as rice, wheat, and sorghum (Ou-Lee et al. 1986), as well as lettuce (Chupeau et al. 1989) and rapeseed (Guerche et al. 1987).

Even more promising is the technique of microprojectile transfer. Klein et al. (1987) have demonstrated that bits of genetic material can be bound to tiny spheres and then shot into intact plant cells. This technique has worked experimentally with corn (Klein et al. 1988) and soybeans (McCabe et al. 1988), and expression of the transferred trait has been demonstrated, though fertile plants have not yet been obtained. Advances with cell culturing techniques, allowing the regeneration of fertile transgenic plants from protoplasts, also suggest that commercial development of grain crops is progressing rapidly (Prioli and Sondahl 1989, Shillito et al. 1989).

Although many traits have been suggested for transfer, such as drought tolerance and nitrogen fixation ability (Barton and Brill 1983, Office of Technology Assessment 1988), only single-gene traits have been transferred into the initial varieties. Many of these traits involve resistance to viruses (Haekema 1989, Powel Abel et al. 1986) or herbicides (Botterman and Leemans 1988), such as glyphosate (Monsanto), bromoxynil (Calgene), or atrazine (Ciba-Geigy).

A gene coding for a toxin, delta endotoxin, which is produced by the bacterium *Bacillus thuringiensis*, has also been transferred into tobacco, tomato, and potato (Leemans 1988). Although the endotoxin (also called Bt toxin) is best known as a defense against lepidoptera, other isolates of *B. thuringiensis* have been obtained that show activity against coleoptera

(Höfte et al. 1987, McPherson et al. 1988) and diptera (Gingrich and El-Abbassi 1988). It is possible that crop varieties may be developed that express multiple forms of endotoxins and thereby gain resistance to beetles, flies, and lepidoptera.

Potential Environmental Impacts

What might be the environmental consequences of widespread acceptance of new varieties containing traits that confer environmental tolerance or pest resistance? Plants with these traits, acquired either by the escape of the engineered crop or from wild plants that have crossed with the crop, could have serious impacts on man-made plant communities.

The most obvious threat comes from plants that might become more serious weeds in agricultural systems. For example, herbicides considered environmentally safe would no longer be effective against weeds that had captured a gene for herbicide resistance, forcing the use of more dangerous chemicals. Similarly, widespread use of insect-resistance traits may lead to the rapid evolution of pest species for which alternate resistance factors or chemical controls would become necessary (Gould 1988).

Impact on natural communities would be less predictable but might have important consequences. For example, weeds with a novel trait that makes them less susceptible to their usual herbivores (such as weedy tomatoes that express Bt toxin) might exhibit greater reproductive success. Rare plant species might be eliminated through competitive displacement, and the genetic diversity of conspecifics might be reduced. Insect communities might also be affected. If lepidopteran herbivores were removed from a plant species, other insects might experience competitive release and become more common. If these released insects are generalist feeders, insects eating other plant species may face greater competition.

Plant communities distant from the site of the novel weed might also be affected. Janzen (1987)[2] notes that some lepidoptera from Guanacaste

The most widely planted crops in the world. Data from Food and Agriculture Organization (1986).	
Crop	Area planted (in 1000 ha.)
1. Wheat	229,000
2. Rice	145,358
3. Corn	131,475
4. Barley	79,645
5. Soybeans	52,103
6. Sorghum	46,807
7. Millet	39,941
8. Cotton	32,279
9. Dry beans	26,207
10. Oats	25,563
11. Potatoes	20,046
12. Peanuts	19,681
13. Sugarcane	15,920
14. Rye	15,492
15. Sunflowers	15,314
16. Rapeseed	14,866
17. Cassava	14,219
18. Coffee	10,498
19. Chick peas	10,456
20. Dry Peas	9,586
21. Grapes	9,046
22. Sugar beets	8,746
23. Sweet potatoes	7,428
24. Sesame	7,079
25. Cacao	5,371
26. Linseed	4,939
27. Tobacco	4,183
28. Broad beans	3,182
29. Lentils	2,882
30. Jute	2,636
31. Tomatoes	2,592
32. Tea	2,537
33. Yams	2,475
34. Watermelons	1,861
35. Onions	1,753

National Park in northwestern Costa Rica cross the central mountain range to spend the dry season in the San Carlos Valley, an area that is becoming increasingly converted to agriculture. The extent to which the migrating species use areas that contain the engineered crop plant or weed could affect insect population dynamics in the distant natural community. In addition, although lepidoptera are herbivores as larvae, many are important pollinators of wild species as adults. Therefore, if larval mortality is

[2]*Also D. Janzen, 1990, personal communication University of Pennsylvania, Philadelphia.*

high, plant communities might also be disrupted by the decreased availability of pollinators.[3] Further disruption of the pollinator and plant communities could occur if the toxin is expressed in plant nectar. To my knowledge, toxin expression in plant nectar has not been tested.

Genetically engineered food crops, however, are not the only plants that present potential environmental risks (Raffa 1989). Genetic manipulation of important commercial forest species, such as pines and poplars, which are

---

Ecological risk analysis must be part of the protocol for releasing new genetically engineered varieties.

---

wind pollinated and disperse pollen and seeds over relatively long distances (Lanner 1966), could have an even greater potential to disrupt natural community dynamics.

Research Agenda

Because the crop varieties first developed using genetic engineering have no widespread close relatives in the United States (Ellstrand and Hoffman 1990), US scientists have been slow to recognize the risk of transfer of genetic material through hybridization (horizontal transfer). However, a recent publication of the Ecological Society of America (Tiedje et al. 1989) describes possible environmental consequences of this type of transfer, whether the organisms are microbes or plants.

In the United Kingdom, the Imperial College and the Institute of Plant Science Research in Cambridge are directing a project on the horizon-

tal transfer of plant genes, funded by the UK Department of Trade and Industry and a consortium of plant biotechnology firms.[4] The first organisms to be studied—sugar beet (*Beta vulgaris*), rapeseed (*Brassica napus*), and potato (*Solanum tuberosum*)—will be transgenic for one or more traits, such as resistance to kanamycin (an antibiotic) or an herbicide. Genetically modified and unmodified plants will be introduced into several habitats at four sites in Britain that represent a range of climatic and soil conditions. Paternity analysis will be used to estimate pollen flow into the control plant population.

Expertise in several disciplines will be useful in assessing the risk posed by transgenic plants worldwide. Systematists who study the relationships between wild and cultivated plants, using techniques such as electrophoresis, chloroplast DNA analysis, and experimental hybridization, may already have evidence suggesting risks of horizontal transfer. Field ecologists, using the statistical analytical methods associated with paternity analysis (Devlin and Ellstrand in press), can design field experiments to determine those factors that may increase risk. Geneticists can build on the information gained in these other areas to estimate selection coefficients and likely persistence of the transgenic plant in a natural community.

Ecological risk analysis must be part of the protocol for releasing new genetically engineered varieties. It will not be sufficient to perform tests in a single environment, because field size, proximity to wild relatives, and environmental conditions are all likely to affect risk.

Political and equity considerations will also be important as companies try to market new varieties in less-developed countries. For example, although increased yields could positively influence balance of payments and social stability in the short term, if novel weeds become a serious problem, then the long-term cost could

outweigh the benefits. Still, an unwillingness by developed countries to enter these markets could be viewed as another form of imperialism, where the raw material of genetic improvement originates in a less-developed country but its benefits are not transferred back.

Both scientific and social analyses are critical for realizing the potential benefits of these new technologies. I hope that both investigations and debate will be prompted by the articles presented in this issue of *BioScience*.

Acknowledgments

I thank Ron Carroll and Fred Gould for critical comments on this manuscript.

References Cited

Barton, K. A., and W. J. Brill. 1983. Prospects in plant genetic engineering. *Science* 219: 671–675.

Brill, W. 1987. Safety issues in perspective. Pages 297–302 in H. M. LeBaron, R. O. Mumma, R. C. Honeycutt, and J. H. Duesing, eds. *Biotechnology in Agricultural Chemistry*. American Chemical Society, Washington, DC.

Botterman, J., and J. Leemans. 1988. Engineering herbicide resistance. *Trends Genet.* 4: 219–222.

Chupeau, M. C., C. Bellini, P. Guerche, B. Maissonneuve, G. Vastra, and Y. Chupeau. 1989. Transgenic plants of lettuce (*Lactuca sativa*) through direct transformation after electroporation of protoplasts. *Bio/Technology* 7: 503–508.

Devlin, B., and N. C. Ellstrand. In press. The development and application of a refined method for estimating gene flow from angiosperm paternity analysis. *Evolution*.

Doebley, J. 1990. Molecular evidence for gene flow among *Zea* species. *BioScience* 40: 443–448.

Ellstrand, N. C., and C. A. Hoffman. 1990. Hybridization as an avenue of escape for engineered genes. *BioScience* 40: 438–442.

Food and Agriculture Organization. 1986. *FAO Production Yearbook 1986*. vol. 40. FAO Statistics Series No. 76, Food and Agriculture Organization, Rome.

Gingrich, R. E., and T. S. El-Abbassi. 1988. Diversity among *Bacillus thuringiensis* active against the Mediterranean fruit fly. Pages 77–84 in *Modern Insect Control: Nuclear Techniques and Biotechnology*. International Atomic Energy Agency, Vienna.

Gould, F. 1988. Evolutionary biology and genetically engineered crops. *BioScience* 38: 26–33.

---

[3]*Although the Bt toxin has been in use for years in commercial agriculture as a topical application (trade names: Dipel, Larvatrol, Thuricide, and others), it exerts short-term selective pressure against insect communities because it degrades rapidly when exposed to field conditions (Thomson 1985). Bt toxins contained in plant tissues are shielded against environmental degradation and would exert constant selective pressures against the susceptible insect community.*

---

[4]*M. Crawley, 1990, personal communication. Imperial College, London, UK.*

Grierson, D., and S. Covey. 1984. *Plant Molecular Biology*. Chapman and Hall, New York.

Guerche, P., M. Charbonnier, L. Jouanin, C. Tourneur, J. Paszkowski, and G. Pelletier. 1987. Direct gene transfer by electroporation in *Brassica napus*. *Plant Sci*. 52: 111–116.

Haekema, A., M. J. Hussman, L. Molendijk, P. M. J. van den Elzen, and B. J. C. Cornelissen. 1989. The genetic engineering of two commercial potato cultivars for resistance to potato virus X. *Bio/Technology* 7: 273–278.

Harlan, J. R., and J. M. J. DeWet. 1972. A simplified classification of cultivated sorghum. *Crop. Sci*. 12: 172–176.

Höfte, H., J. Seurnick, A. Van Houtven, and M. Vaeck. 1987. Nucleotide sequence of a gene encoding an insecticidal protein of *Bacillus thuringiensis* var. *tenebrionis* toxic against Coleoptera. *Nucleic Acids Res*. 15: 7183–7186.

Janzen, D. J. 1987. Insect diversity of a Costa Rican dry forest: why keep it and how? *Biol. J. Linn. Soc*. 30: 343–356.

Klein, T. M., M. Fromm, A. Weissinger, D. Tomes, S. Schaaf, M. Sletten, and J. C. Sanford. 1988. Transfer of foreign genes into intact maize cells with high-velocity microprojectiles. *Proc. Natl. Acad. Sci*. 85: 4305–4309.

Klein, T. M., E. D. Wolf, R. Wu, and J. C. Sanford. 1987. High-velocity microprojectiles for delivering nucleic acids into living cells. *Nature* 327: 70–73.

Lanner, R. M. 1966. Needed: a new approach to the study of pollen dispersion. *Silvae Genet*. 15: 50–52.

Leemans, J. 1988. Engineering insect and herbicide-resistant crops. Pages 77–81 in R. T. Fraley, N. M. Frey, and J. Schell, eds. *Genetic Improvement of Agriculturally Important Crops*. Cold Spring Harbor Laboratory, Cold Spring Harbor, NY.

McCabe, D. E., W. F. Swain, B. J. Martinell, and P. Christou. 1988. Stable transformation of soybean (*Glycine max*) by particle acceleration. *Bio/Technology* 6: 923–926.

McPherson, S. A., F. J. Perlak, R. L. Fuchs, P. G. Marrone, P. B. Lavrik, and D. A. Fischhoff. 1988. Characterization of the coleopteran-specific protein gene of *Bacillus thuringiensis* var. *tenebrionis*. *Bio/Technology* 6: 61–66.

Office of Technology Assessment. 1988. *Field Testing Engineered Organisms: Genetic and Ecological Issues. New Developments in Biotechnology 3*. United States Congress, Washington, DC.

Ou-Lee, T.-M., R. Turgeon, and R. Wu. 1986. Expression of a foreign gene linked to either a plant-virus or a *Drosophila* promoter, after electroporation of protoplasts of rice, wheat, and sorghum. *Proc. Natl. Acad. Sci*. 83: 6815–6819.

Powel Abel, P., R. S. Nelson, B. De, N. Hoffmann, S. G. Rogers, R. T. Fraley, and R. N. Beachy. 1986. Delay of disease development in transgenic plants that express tobacco mosaic virus coat protein gene. *Science* 232: 738–743.

Prioli, C. M., and M. R. Sondahl. 1989. Plant regeneration and recovery of fertile plants from the protoplasts of maize (*Zea mays* L.). *Bio/Technology* 7: 589–594.

Raffa, K. F. 1989. Genetic engineering of trees to enhance resistance to insects. *BioScience* 39: 524–534.

Raineri, D. M., P. Bottino, M. P. Gordon, and E. W. Nester. 1990. *Agrobacterium*-mediated transformation of rice (*Oryza sativa* L.). *Bio/Technology* 8: 33–38.

Shillito, R. D., G. K. Carswell, C. M. Johnson, J. J. DiMairo, and C. T. Haims. 1989. Regeneration of fertile plants from protoplasts of elite inbred maize. *Bio/Technology* 7: 581–587.

Strange, C. 1990. Cereal progress via biotechnology. *BioScience* 40: 5–9, 14.

Thomson, W. T. 1985. *Agricultural Chemicals. book 1. Insecticides, Acaricides, and Ovicides*. Thomson, Fresno, CA.

Tiedje, J. M., R. K. Colwell, Y. L. Grossman, R. E. Hodson, R. E. Lenski, R. N. Mack, and P. J. Regal. 1989. The planned introduction of genetically engineered organisms: ecological considerations and recommendations. *Ecology* 70: 298–315.

Wilson, H. W. 1990. Gene flow in squash species. *BioScience* 40: 449–455.

THOUGHT EVOKERS

- Describe the potential environmental impacts that might be produced by genetic engineering of crop plants.

- Explain the prevailing techniques used in genetic engineering of crop plants.

VI

Animal Structure and Function

Congruent with human population growth is the tremendous increase in the release of pollutants into our environment. A recent report by the National Wildlife Federation stated that more than 4.9 billion tons of toxic chemicals were released into the air, water, and soil of the United States in 1990. These chemicals, including heavy metals, radioactive compounds, and biocides of all types, are capable of causing harmful effects—from cancer to birth defects—for all living systems. One of these metals, iron, has recently been implicated as an underlying cause of heart disease and, according to a recent Finnish study, may in fact rank second behind smoking as a major causative agent of cardiac problems.

Evidence continues to build regarding the dysfunction of human physiological processes caused by a continuing, and increasing, use of steroids. The article "Pumped Up" points out that "With the goal of being bigger, stronger, and faster, many young Americans are playing a risky game of chemical roulette. Their credo: 'Die young, die strong.'" As research concerning steroid use suggests, their credo may be fulfilled.

Recent botanical studies of various compounds found in plants have contributed to a major breakthrough regarding the use of taxol, a compound derived from the bark of yew trees, in the treatment of some forms of human cancers. Discovering plant derivatives that may prove useful in the treatment of various human conditions furnishes further evidence for maintaining the biotic diversity of all ecosystems.

And the earth's population? Even with scientific advances in birth control, the numbers increase. Today over 5.5 billion people inhabit the planet, and many biologists maintain that we have exceeded the earth's carrying capacity. If we and our children are to experience the quality of life that we all seek, then we must bring our numbers under control. In the United States, teenage pregnancies alone result in over one million offspring yearly, and the numbers are expected to rise dramatically.

17

Iron and Your Heart

By Steven Findlay, Doug Podolsky, and Joanne Silberner

Well, blow me down. Popeye's habit of downing iron-rich spinach might make him a candidate for Toonland General's cardiac-care unit. That is, *if* surprising new statistical findings are confirmed that link even normal levels of iron in the blood to a higher risk of heart attack.

The provocative study by Finnish researchers, published last week in the American Heart Association's scientific journal *Circulation*, concluded that the amount of stored iron in the body ranks second only to smoking as the strongest risk factor for heart disease and heart attacks. U.S. heart experts last week eyed the study's findings with steely skepticism. Yet researchers at prestigious institutes are so intrigued that they are quickly moving to reanalyze their heart-disease data to see whether the Finnish findings hold up in the United States.

Experts acknowledge that the study, though involving only 1,900 men from eastern Finland, was carefully done. A team of epidemiologists led by Jukka Salonen of the University of Kuopio tested healthy Finns ages 42 to 60 for serum levels of ferritin, an iron-storing protein. Salonen's team then followed the patients for five years. By that time, 51 had suffered heart attacks. Analysis showed that the men with more than 200 micrograms of ferritin per liter of blood—a level generally considered normal—were more than twice as likely to have heart attacks compared with men with levels below 200. The risk of heart attack more than quadrupled in men with high levels both of iron and of low-density lipoproteins, or LDLs—cholesterol-carrying packages of fats and proteins.

Even as experts caution that much remains to be learned before new diets are urged, the implications have scientists buzzing. The study represents the first hard clinical evidence for a largely ignored 11-year-old theory of heart disease. The central tenet is that high iron levels add to the risk of heart disease and that low iron levels protect against it. Salonen and others now propose that iron elevates the risk of heart attack by promoting chemical reactions between LDLs and oxygen. That kicks off a cascade of events that narrow the heart's arteries, so that even a small clot can block blood flow and cause a heart attack. And iron seems to feed the formation of free radicals, unstable oxygenated molecules that have been implicated not only in damage to heart muscle after a heart attack but also in cancer, diabetes, arthritis, and the aging process.

Fallen angel? The iron theory slashes through some of heart disease's thorniest underbrush. It suggests why heart disease is so rare among premenopausal women (women lose iron during menstruation). It offers a plausible explanation about cross-cultural differences in heart-disease risk (societies that consume more red meat are at greater risk because of meat's high iron content). And it may explain why fish oil protects against heart disease (it prolongs bleeding so that iron stores are depleted); why aspirin protects against heart attack (it can cause small amounts of internal bleeding); and even why oral contraceptives raise the heart-disease risk for older female smokers (the pill reduces menstrual blood loss and promotes clotting).

> ## An Enemies List
>
> Behind a Finnish study is the message that iron might rank No. 2 as a cause of heart attacks.
>
> THE TOP RISK FACTORS:
> 1. **Smoking**
> 2. **Iron**
> 3. **High cholesterol**
> 4. **High blood pressure**
> 5. **Diabetes**
> 6. **Family history**
>
> *USN&WR*—Basic data: American Heart Association; National Heart, Lung and Blood Institute; *Circulation*

If follow-up studies buttress the theory, iron could become the fallen angel of nutrition. It has long been recognized as an essential nutrient. That has led to its incorporation in dietary supplements, some of which deliver 400 to 700 percent of the 18-milligram recommended daily allowance. Many Americans were reared on ads touting iron-rich Geritol as the cure for "tired blood." (Doug Cox, a spokesman for Geritol maker Smith-Kline Beecham, notes that the company no longer makes such claims and labels the Finnish study "preliminary.")

The cardiology community is unlikely to accept the hypothesis until more scientists weigh in. Any single study can be confounded by unexpected factors. The high iron levels may be the result of an underlying disorder that is the real cause of heart disease. The Finnish subjects may carry a gene that makes them particularly sensitive to the effects of iron. (One reason for chief researcher

Salonen's work is that he himself has hemochromatosis, a genetic condition in which the body absorbs and stores excessive iron.) It is possible that charting high iron levels could simply net people who eat a lot of red meat or have other habits that put them at risk. And the researchers only checked middle-aged white men; whether the iron connection will hold in women as well as for other age groups and races remains to be seen.

Long knives. Some researchers already dispute some of the iron theory's assertions. William Castelli, head of a large heart study of the residents of Framingham, Mass., notes that while he expects the iron connection will pan out, he doesn't buy the hypothesis as an explanation for the relative rarity of heart disease in premenopausal women. Because of studies in Framingham and elsewhere showing that women who take estrogen after menopause have a lower incidence of heart disease, Castelli

believes that protection comes from high estrogen, not low iron.

Answers may come rapidly. Charles Hennekens, an epidemiologist at Harvard Medical School and Brigham and Women's Hospital in Boston,

---

About 20 percent of the iron Americans consume comes from fortified foods.

---

plans to check iron levels in stored blood samples from 22,000 doctors whose heart-attack risks he has studied since 1982. The results could be ready in three months, he says, and he hopes to repeat the study in women as well. Even if such efforts confirm the connection, further research will be needed to determine if lowering iron levels will actually prevent heart disease.

The findings are in fact just the latest in a string of developments that

put the metal at the center of controversy. The iron-heart theory met dead silence when Jerome Sullivan, then a Florida pathologist in training and now director of clinical laboratories at the Veterans Affairs Medical Center in Charleston, S.C., first published it in 1981. Sullivan was so convinced he was right that for years he controlled his own iron levels by donating blood every month or so. Now, after a suspected bout with cytomegalovirus, an infection that could make his blood unsafe for others, he refrains from donating. Instead, he has his lab techs practice bloodletting on him. "I try to do it as often as I can," Sullivan says.

Of late, scientists and health officials have increasingly been rethinking just how much dietary iron is good for people. And a consensus seems to be forming that calls into question the generally accepted notion that an iron-rich diet is beneficial. Iron's link to cancer is being investigated, for example. A 1989

Now, the testing question

Concerns about iron's role in heart disease, if ultimately confirmed, surely will touch off a debate over iron tests. Who should be tested? And should testing be a matter of public health policy? Even before the Finnish study appeared, some proponents of the iron theory were advising anyone age 30 or over to have a blood test for iron every five years.

The principal virtue of such broad screening would be to help ferret out many of the estimated 32 million Americans who carry a faulty gene that prompts their body to harbor too much iron. There is no test for the gene itself, but anyone with a high blood level of iron is a suspect. Siblings and children would be candidates for testing as well, since the condition is inherited.

Normal numbers

Two widely used blood tests for iron generally cost from $25 to $75 each. The first measures ferritin, a key iron-storing protein. Typical levels, expressed in micrograms per liter of blood, range from 100 to 150 in men and from 50 to 100 in women. Above 300 is usually considered very high.

The results should always be discussed with a doctor, since some physicians take recent studies seriously enough to worry about even moderately elevated ferritin levels. Moreover, the test might bear repeating. Erroneously high readings are relatively

common, especially in people with health problems like bowel and liver diseases, arthritis, and certain forms of cancer. Oral contraceptives and estrogen pills can also elevate readings.

The second test measures the blood's ability to carry iron, termed total iron-binding capacity and abbreviated on laboratory forms as TIBC. It too is expressed in micrograms per liter of blood. Anyone with a family history of hemochromatosis, a condition involving a toxic buildup of iron in the body, should get a TIBC test as well as a separate check of a blood protein called transferrin. The "transferrin saturation" in the blood is expressed as a percentage. Normal is around 33 percent; a reading higher than 62 percent for men and 50 percent for women usually signals hemochromatosis. Results between 40 and 60 percent may indicate that a single copy of the hemochromatosis gene is present.

None of these tests checks specifically for iron-deficiency anemia. That diagnosis requires measuring the number of grams of hemoglobin, the iron-based molecule that carries oxygen, per liter of blood, as well as hematocrit, the percentage of red blood cells in the blood. Both are sensitive tests for anemia. Some experts advise anyone diagnosed as anemic to get a ferritin test as well, to more precisely establish the body's store of iron so that it can be tracked over time. And those with persistently low levels of ferritin should have their hemoglobin measured for anemia.

study of 14,000 adults funded by the National Institutes of Health found that men who had high blood levels of iron were 37 percent more likely to develop cancer of the colon, lung, bladder, and esophagus than men who had the lowest iron levels. A study among enrollees in Kaiser Permanente's health plan in California yielded a similar finding.

Radical notion. The ways in which iron may trigger artery clogging and heart disease also appear to be at work not only in cancer but in arthritis and the aging process. Although the evidence is still largely circumstantial, belief is growing that the common thread is iron's complicity with oxygen in forming unstable free radicals that disrupt chemical bonds and can damage DNA molecules. The newest theories suggest that free-radical damage, triggered by excessive iron in the blood, builds over time. This might induce cancer, break down tissues, and impair organ function.

Yet iron *deficiency* is still an important, if waning, public-health prob-

---

Up to 1 in 10 American adults may carry a gene that predisposes them to store excessive iron.

---

lem. From 10 to 20 percent of infants and 8 percent of women in low-income families have iron-poor diets. Menstruating and pregnant women in all income and age brackets are also prone to iron deficiency. Because the condition can derail a healthy pregnancy, doctors commonly prescribe iron supplements for pregnant women.

The lingering prevalence of iron deficiency has preserved the government's 50-year-old requirement that processed breads, flour, and other grain products be "fortified" with iron. Between 10 and 20 percent of the average American's dietary iron comes from fortified foods. The practice began during World War II when it was discovered that milling removed wheat's natural iron. For most Americans, however, the practice is no longer necessary and may even be harmful, since the vast majority of Americans eat a diet that contains more than enough iron, claims Randall Lauffer, a Harvard University biochemist, author of "Iron Balance" (St. Martin's Press, 1991, $19.95), and editor of the textbook "Iron and Human Disease" (CRC Press, 1992). "Fortification," says Lauffer, "only adds to the chance that some people

---

Seeking Other Villains

Iron is hardly the only suspect in the search for heart-disease culprits. In previous studies, the same Finnish researchers who implicated iron found that high blood levels of copper and low levels of selenium often precede heart attacks. In the usual scenario, moreover, heart attacks occur when a narrowed artery in the heart is blocked by a blood clot; scientists are hunting for substances that lead to the narrowing, as well as others that promote clotting. Among the hot prospects:

Lipoprotein(a)

This molecular package of cholesterol, fat, and protein is a variation on LDL, or low-density lipoprotein (the so-called bad cholesterol). Epidemiological studies link high levels of lipoprotein(a) to early heart attacks that are otherwise mysterious. The molecule may be a double threat; evidence suggests it can both dump its cholesterol into artery walls and promote the formation of clots.

Viruses and bacteria

Remnants of herpes viruses sometimes show up in swollen arterial walls; figuring out whether that is a cause or an effect of atherosclerosis has kept researchers here and in Europe busy for years. Recent studies from Washington and Finland indicate that infection with one common strain of pneumonia-causing bacterium doubles or triples the risk of a heart attack.

Small, dense LDL

This form of LDL was given its uncomplimentary name because its components are more tightly packed than in other forms of LDL. Inheriting it may set people up for a heart attack, but just how remains a mystery. Researchers at the University of California at Berkeley estimate that a quarter of the U.S. population may be genetically predisposed to produce high levels of small, dense LDL.

Albumin

Several groups of researchers have shown that people with low blood levels of this protein—a major component of egg white—are more prone to heart disease. Low albumin, contends University of Pittsburgh epidemiologist Lewis Kuller, is as strong a predictor as age, blood pressure, cigarette smoking, and cholesterol level.

Homocysteine

A buildup in the blood of homocysteine, one of the building blocks of protein, irritates the delicate lining of the arteries, promotes clot formation, and activates LDL into a form that lodges in arteries. Only 1 to 2 percent of people have even moderately high levels of homocysteine, but in a small study from Ireland published last year, about 30 percent of people under the age of 55 with clogged arteries had high levels of homocysteine; all those tested with clear arteries had normal levels.

Given the complexity of the heart's plumbing, most scientists expect to see more than one suspect convicted. Physics may yield a grand unified theory one of these days, but biology is unlikely to be so accommodating.

will be eating a dangerously excessive amount of iron."

What's too much? Adding to the controversy is the practice of cereal makers of adding the entire 18-mg recommended daily allowance of iron to cereals like Product 19 and Total. Says Lauffer, "They are capitalizing on the false notion that the more iron you eat the better." A spokesperson for Kellogg says the company's nutritionists and medical experts advise that "we wait for more research."

While iron deficiency is well understood and sometimes overdiagnosed, the new concept of iron overload has yet to be defined. Lauffer believes that half the U.S. population suffers from potentially toxic levels of iron, but no scientist has verified his assessment. The only large-scale study of iron levels in the general population, done between 1976 and 1980, found that 10 percent of men ages 20 and above have blood ferritin levels greater than 200 micrograms per liter.

There's also no clear agreement on the danger level. Average levels, based on ferritin measurements, are

---

Donating blood can cut men's and menopausal women's iron to the level of a menstruating woman's.

---

100 to 150 micrograms per liter in adult men and 50 to 100 in premenopausal adult women. The Finnish study signaled danger at a level of 200 micrograms or above; the cancer study used a different measure of iron in the body but also found that levels at the high end of the normal range were linked with a heightened risk of the disease. Some doctors diagnose iron overload at ferritin levels of 300, while others consider that within normal bounds. The uncertainty echoes the long debate in the 1970s and early 1980s over critical numbers for cholesterol and blood pressure.

Much of what is known about the effects of too much iron on the body has been gleaned from the experience of people with hemochromatosis. About 1 million Americans, or 1 person in 250, have the inherited disorder. Over a lifetime, victims' iron stores build up to dangerous levels. Without frequent bloodletting to reduce iron stores, they face diabetes, liver disease, and damage to the joints and heart, usually starting in their 40s and 50s. Untreated cases are usually fatal. No genetic test exists, so the condition often goes undiagnosed until the victims, mostly men, start to get sick—and then their symptoms, including fatigue, weakness, aching joints, and loss of sexual desire, are often mistaken for something else for months or even years.

More than people with full-blown hemochromatosis are at risk. In recent years, geneticists have confirmed that as many as 32 million Americans carry one copy of the hemochromatosis gene. (Sufferers carry two copies, but not everyone with two copies develops the disease.) Evidence is mounting that carrying just one copy can lead to absorption of too much dietary iron, but not as much as for people carrying two genes. The single-gene carriers could be prime candidates for a heart attack if the iron-heart theory is verified. Speculates James Cook, an iron expert and director of hematology at the University of Kansas Medical Center: "If this problem does exist, people with the single gene could account for most of those at risk in our society."

The diet question. Some experts assert that iron fortification of foods and iron in vitamin pills endanger this relatively large segment of the population and the extra iron should therefore be removed. Researchers reporting in the *Journal of the American Dietetic Association* last February, however, said there is no evidence that iron supplements or iron in foods have increased the incidence of hemochromatosis disease.

In fact, how dietary iron figures into iron overload is poorly understood. The Finnish study, for instance, did not focus on the effect of diet on blood levels of iron. In hemochromatosis patients and perhaps those with a single copy of the gene, the relationship is direct; the more iron these people eat, the faster iron accumulates. But without the faulty gene, experts say, an iron-rich diet won't necessarily lead to iron overload. For most people, says Cook, there is no clear correlation between the iron in their diet—unless it is outrageously high—and the iron level in their blood. Only people who are iron-deficient, he claims, will start absorbing more iron from their food.

Author Lauffer counters that circumstantial but strong evidence suggests that what you eat does affect levels of body iron. Studies show, he says, that people in countries with iron-rich diets have higher average blood levels of iron than people in countries with low-iron diets, vegetarians also usually have lower iron levels than meat eaters—and a lower risk of heart disease. "I believe that over years and years iron can build up in everyone," says Lauffer. If so, meat is the major suspect. Studies show clearly that 20 to 30 percent of the iron in beef, poultry, and seafood is absorbed by the body, compared to 3 to 5 percent of the iron from vegetables and legumes.

For worrywarts, the drastic solution is to donate blood periodically to get rid of some iron. It's also socially beneficial. But even iron theorists like Sullivan, a former blood donor himself, agree that such a strategy should be proved effective before it becomes public health policy. Healthy people taking iron supplements can stop, not necessarily because of the Finnish study but because doctors and nutritionists have suggested for years that people not dose themselves unless they've been diagnosed with anemia or are pregnant. Cast-iron cookware leaches only small amounts of iron into food. And municipal tap water has little iron. Well water, particularly in north central states, may contain 10 to 50 milligrams of iron per liter, according to the National Ground Water Association in Dublin, Ohio. Ground water contractors and some state health departments can test samples, and water-treatment specialists can install filters to remove excess iron.

It's far too early in the game to recommend universal dietary changes. But people who insist on hedging their bets can take the advice of Marion Nestle, head of the nutrition department at New York University. "Whether the iron theory is correct or not," she says, "if you're eating a diet that's low in fat and high in fruits, vegetables and grains, you probably don't have to worry about it."

THOUGHT EVOKERS

- Explain why high iron levels in the blood may add to the risk of heart disease.

- Why is it suggested that premenopausal women may be at lower risk of heart disease than postmenopausal women?

- Discuss the suggested methods by which iron might contribute to arterial clogging.

18 Pumped Up

With the goal of being bigger, stronger, and faster, many American teenagers are playing a risky game of chemical roulette. Their credo: Die young, die strong.

By Joannie M. Schrof

It's a dangerous combination of culture and chemistry. Inspired by cinematic images of the Terminator and Rambo and the pumped-up paychecks of athletic heroes with stunning physiques and awesome strength, teenagers across America are pursuing dreams of brawn through a pharmacopoeia of pills, powders, oils, and serums that are readily available—but often damaging. Despite the warnings of such fallen stars as Lyle Alzado, the former football player who died two weeks ago of a rare brain cancer he attributed to steroid use, a *U.S. News* investigation has found a vast teenage subculture driven by an obsession with size and bodybuilding drugs. Consider:

• An estimated 1 million Americans, half of them adolescents, use black-market steroids. Countless others are choosing from among more than 100 other substances, legal and illegal, touted as physique boosters and performance enhancers.

• Over half the teens who use steroids start before age 16, sometimes with the encouragement of their parents. In one study, 7 percent said they first took "juice" by age 10.

• Many of the 6 to 12 percent of boys who use steroids want to be sports champions, but over one third aren't even on a high-school team. The typical user is middle-class and white.

• Fifty-seven percent of teen users say they were influenced by the dozen or so muscle magazines that today reach a readership of at least 7 million; 42 percent said they were swayed by famous athletes who they were convinced took steroids.

• The black-market network for performance enhancers is enormous, topping $400 million in the sale of steroids alone, according to the U.S. Drug Enforcement Administration. Government officials estimate that there are some 10,000 outlets for the drugs—mostly contacts made at local gyms—and mail-order forms from Europe, Canada, and Mexico can be found anywhere teenagers hang out.

• The nation's steroid experts signaled the state of alarm when they convened in April in Kansas City to plan the first nationwide education effort.

• Even Arnold Schwarzenegger, who has previously been reluctant to comment on his own early steroid use, has been prompted to speak out vigorously about the problem. The bodybuilder and movie star is the chairman of the President's Council on Physical Fitness and Sports.

One 13-year-old who had taken steroids for two years stopped growing at 5 feet. "I get side effects," says another teen who has used steroids for three years. "But I don't mind; it lets me know the stuff is working."

Performance drugs have an ancient history. Greek Olympians used strychnine and hallucinogenic mushrooms to psych up for an event. In 1886, a French cyclist was the first athlete known to die from performance drugs—a mixture of cocaine and heroin called "speedballs." In the 1920s, physicians inserted slices of monkey testicles into male athletes to boost vitality, and in the '30s, Hitler allegedly administered the hormone testosterone to himself and his troops to increase aggressiveness.

The use of anabolic steroids by weight lifters in the Eastern bloc dates back at least to the 1950s, and the practice has been spreading ever since among the world's elite athletes. But recent sensations in the sports world—Ben Johnson's record-shattering sprints at the Seoul Olympics and the signing of Brian Bosworth to the largest National Football League rookie contract ever *after* he tested positive for steroids—have attracted both young adults and kids to performance enhancers like never before, say leading steroid experts. These synthetic heroes are revered rather than disparaged in amateur gyms around the country, where wannabe Schwarzeneggers rationalize away health risks associated with performance-enhancing drugs.

Weighing in. The risks are considerable. Steroids are derivatives of the male hormone testosterone, and although they have legitimate medical uses—treatment of some cancers, for example—young bodybuilders who use them to promote tissue growth and endure arduous workouts routinely flood their bodies with 100 times the testosterone they produce naturally. The massive doses, medical experts say, affect not only the muscles but also the sex organs and nervous system, including the brain. "Even a brief period of abuse could have lasting effects on a child whose body and brain chemistry are still developing," warns Neil Carolan, who directs chemical dependency programs at BryLin Hospitals in Buffalo and has counseled over 200 steroid users.

Male users—by far the majority—can suffer severe acne, early balding, yellowing of the skin and eyes, development of female-type breasts, and shrinking of the testicles. (In young boys, steroids can have the opposite effect of painfully enlarging the sex organs.) In females, the voice deepens permanently, breasts shrink, periods become irregular, the clitoris swells, and hair is lost from the head but grows on the face and body. Teen users also risk stunting their growth, since steroids can cause bone growth plates to seal. One 13-year-old who had taken steroids for two years stopped growing at 5 feet. "I get side effects," says another teen who has used steroids for three years. "But I don't mind; it lets me know the stuff is working."

In addition to its physical dangers, steroid use can lead to a vicious cycle of dependency. Users commonly take the drugs in "cycles" that last from four to 18 weeks, followed by a lengthy break. But during "off" time, users typically shrink up, a phenomenon so abhorrent to those obsessed with size that many panic, turning back to the drugs in even larger doses. Most users "stack" the drugs, taking a combination of three to five pills and injectables at once; some report taking as many as 14 drugs simultaneously. Among the most commonly used are Dianabol ("D-Ball"), Anavar, and Winstrol-V, the same type of steroid Ben Johnson tested positive for in 1988. "You wouldn't believe how much some guys go nuts on the stuff," says one teen bodybuilder from the Northeast. "They turn into walking, talking pharmacies."

Despite massive weight gains and sharply chiseled muscles, many steroid users are never quite happy with their physiques—a condition some researchers have labeled "reverse anorexia." "I've seen a kid gain 100 pounds in 14 months and still not be satisfied with himself," reports Carolan. If users try to stop,

they can fall into deep depressions, and they commonly turn to recreational drugs to lift their spirits. Even during a steroid cycle, many users report frequent use of alcohol and marijuana to mellow out. "I tend to get really depressed when I go off a cycle," says one Maryland teen, just out of high school. "On a bad day, I think, 'Gee, if I were on the stuff this wouldn't be happening.' "

"Juicers" often enjoy a feeling of invincibility and euphoria. But along with the "pump" can come irritability and a sudden urge to fight. So common are these uncontrolled bursts of anger that they have a name in the steroid culture: "roid rages." The aggression can grow to pathological proportions; in a study by Harvard researchers, one eighth of steroid users suffered from "bodybuilder's psychosis," displaying such signs of mental illness as delusions and paranoia. So many steroid abusers are ending up behind bars for violent vandalism, assault, and even murder that defense attorneys in several states now call on steroid experts to testify about the drugs' effects.

What steroids do in the long run is still unknown, largely because not one

federal dollar has been spent on long-term studies. Although Lyle Alzado was convinced that steroids caused his brain cancer, for example, there is no medical evidence to prove or disprove the link. But physicians are concerned about occasional reports of users falling ill with liver and kidney problems or dropping dead at a young age from heart attacks or strokes. Douglas McKeag, a sports physician at Michigan State University, is compiling a registry of steroid-related illnesses and deaths to fill the gaping hole in medical knowledge. McKeag sees preliminary evidence that steroid use might cause problems with blood-cell function that could lead to embolisms in the heart or lungs. "If that turns out to be true," he says, "then bingo—we'll have something deadly to warn kids about."

Dianabol desperadoes. Unfortunately, even that sort of documented health threat is unlikely to sway committed members of the steroid subculture. One widely shared value among users is a profound distrust of the medical community. Their suspicion is not totally unjustified. When steroid use was first becoming popular in the late 1950s, the medical community's response was to claim that they didn't enhance athletic ability—a claim that bulked-up users knew to be false. When that failed to deter users, physicians turned to scare tactics, branding steroids "killer drugs," again without hard evidence to back up the claim. As a result, self-styled "anabolic outlaws" and "Dianabol desperadoes" have sought guidance not from doctors but from the "Underground Steroid Handbook," a widely distributed paperback with detailed instructions for the use of more than 80 performance enhancers. "I know that proper steroid therapy can enhance your health; it has enhanced mine," writes author Daniel Duchaine. "Do you believe someone just because he has an M.D. or Ph.D. stuck onto the end of his name?" Or kids simply make up their own guidelines. "If you take more kinds at once, you get a bigger effect, and it's less dangerous because you're taking less of each kind," reasons one 18-year-old football player who has been taking steroids for two years.

Although even the steroid handbook mentions health risks particular to children and adolescents, in the end most young users seem unfazed by the hazards. In one poll, 82 percent said they didn't believe that steroids were harming them much, and, even more striking, 40 percent said they wouldn't stop in any case. Their motto: "Die young, die strong, Dianabol." The main drawback to steroids, users complain, is that many brands must be administered with huge syringes. The deeper the needle penetrates the muscle, the less juice squandered just under the skin. Inserting the 1 1/2-inch needles into their buttocks or thighs leaves many teens squeamish, and they often rely on trusted friends to do the job. "The first time I tried to inject myself, I almost fainted, and one of my friends did faint," remembers a 19-year-old from Arizona. "Sometimes one of the guys will inject in one side of his butt one day and the other the next. Then, we all laugh at him because he can barely sit down for the next three days."

Local "hard-core" gyms, patronized by serious weight lifters, are the social centers of the steroid culture. Teenagers caught up in the bodybuilding craze—typically white, middle-class suburbanites—commonly spend at least three hours a day almost every day of the week there, sometimes working out in the morning before school and again after school is out. Here they often meet 20-to-30-year-old men using steroids to bulk up for power lifting and bodybuilding shows or members of what steroid experts call the "fighting elite"—firefighters, bouncers, even policemen—synthetically boosting the physical strength they need to do their jobs. "Our role model is this older guy, the biggest guy at the gym," says one 17-year-old. "He's not a nice guy, but he weighs 290 pounds without an ounce of fat . . . that's our goal."

The older steroid veterans not only inspire kids to try the drugs but often act as the youngsters' main source for the chemicals. Sometimes, it's the gym owner who leads kids to a stash of steroids hidden in a back room; sometimes, it's a lifter who keeps the drugs in a dresser drawer at home and slips

kids his phone number. Once in a while, it's a doctor or veterinarian who writes out endless prescriptions for the boys or for an unscrupulous coach. And too often, it's overzealous parents who push the drugs on their children. "My stepdad says he's going to start me up on steroids as soon as I'm done growing," says one freshman who wants to play pro football, "But I think he's just joking." Greg Gaa, director of a Peoria, Ill., sports-medicine clinic, says he has gotten calls from up to a dozen parents a year who want him to supply illegal performance enhancers to their children.

A vast black market across America guarantees kids ready access to steroids in big cities and small towns alike. Typically, the drugs are shipped via private couriers from sources in other countries. Two order forms obtained by *U.S. News* require a minimum order of $75, offer 14 different steroids (ranging from $15 to $120 per bottle), and promise 48-hour delivery

---

Greg Gaa, director of a Peoria, Ill., sports-medicine clinic, says he has gotten calls from up to a dozen parents a year who want him to supply illegal performance enhancers to their children.

---

for an extra $20. Though the order forms, sent out six months apart, are identical and obviously the work of the same operation, the company name and address have been changed, apparently to outsmart investigators. In the earlier mailing, it's Mass Machine, located in Toronto. In the later form, it's Gym Tek Training, located in New Brunswick, Canada. Jack Hook, with the U.S. Drug Enforcement Administration in San Diego, describes a sting operation in which undercover agents from the DEA and the California Bureau of Narcotics posing as bodybuilders met up with a European gym owner and ordered $312,000 worth of steroids; the seller was nabbed in February when the shipment arrived via Federal Express.

Sometimes, kids themselves get into

the act. Twenty-five percent say they sell the drugs to support their expensive habit. One Virginia 12th grader tells of fellow students who stole steroids from a drugstore where they worked and made "a killing" selling them around school. "Everybody knows you just go to this one guy's locker, and he'll fix you up," says the teen. A typical 100-tablet bottle of steroids—a month's supply—usually runs from $80 to $100 on the black market, but naive high schoolers often pay three times that amount.

"The challenge of getting ahold of the stuff is half the fun," admits one 17-year-old from Iowa, who tells of meeting dealers in parking lots and taste-testing drugs that look like fakes. Drug-enforcement agents estimate that 30 to 50 percent of the illegal muscle builders teens buy are phony. One Chicago-area youth spent $3,000 on what turned out to be a saline solution. Investigators have seized pills that turned out to be penicillin—deadly to some—and phony oils that were poorly packaged and rampant with bacteria. In April, two Los Angeles dealers were convicted of selling a counterfeit steroid that caused stomach pain, vomiting, and a drop in blood pressure; the substance was a European veterinary drug used in show animals.

Subbing dangers. Since February 1991, when nonmedical steroid distribution became a federal offense punishable by five years in prison, several drugs touted as steroid alternatives have also flourished underground. The top seller this year is a compound called clenbuterol, which is used by veterinarians in other countries but is not approved for any use in the United States. The drug recently led to problems in Spain, where 135 people who ingested it fell ill with headaches, chills, muscle tremors, and nausea.

Human growth hormone, the steroid alternative Lyle Alzado used during his failed efforts at an NFL comeback, is medically used to treat dwarfism by stimulating growth. Its price, up to $1,500 for a two-week supply, is formidable, yet 5 percent of suburban Chicago 10th-grade boys surveyed in March by Vaughn Rickert of the University of Arkansas for Medical

Sciences claim to have used the hormone. Although the body produces the substance naturally, too much can cause acromegaly, or "Frankenstein's syndrome," which leads to distortion of the face, hands, and feet and eventually kills its victim.

Gamma-hydroxybutyrate (GHB) is a dangerous substance now popular among size seekers because it stimulates the release of human growth hormone. It also leads to comas. One Midwestern teen drank a GHB formula before going out to his high-school prom. He never made it. Within 20 minutes, he fell comatose and was rushed to the hospital to be revived. The Centers for Disease Control reports 80 recent hospitalizations from GHB use.

Many of the steroid alternatives that kids turn to come from an unlikely source: the local health-food store. For years, well-meaning coaches have persuaded kids to stay off steroids by opting for legal (and presumably safe) performance aids advertised ad nauseam in muscle magazines and sold in every shopping mall. Kids, happy to find a legal boost, empty their pockets on colorful packages that can cost up to $200 for a month's supply. But chemicals marketed as dietary supplements—essentially as food—undergo far less scrutiny than those marketed as drugs. "We have virtually no idea what's inside some of these products," warns Food and Drug Administration supplement specialist Don Leggett. "Just the other day someone asked about three new chemical compounds, and we couldn't even identify them. The substances aren't even on the books yet." Not long ago, he points out, clenbuterol and GHB were available in some health stores. Leggett is part of a task force now trying to assess the safety of a dozen common ingredients found in the bulking-up formulas, including chromium, boron, and plant sterols.

Cracking the culture. Meanwhile, the ambiance of gyms and health-food stores serves to cloak the use of performance-enhancing drugs in the veneer of a healthy lifestyle. Since all of the trappings of their world have to do with hard work, fitness, and vitality, kids who use the substances see them as just another training aid, not much different from Gatorade or a big steak dinner. "We're not freaks or addicts," asserts one teen. "We're using modern science to help us reach our goals."

Educators agree that users tend to be mainstream kids. "These kids aren't your typical drug abuser," says Dick Stickle, director of Target, the high-school sports association that hosted a meeting in April of 65 experts who worked to plot a strategy for educating teens about the drugs' risks. "They have goals, they have pride; we've got to play on that pride." The group plans to send a book of guidelines for combatting the use of steroids and other performance enhancers to every secondary school, 37,000 in all, early this fall. But reaching secondary schools may not be enough: A Peoria, Ill., teacher was recently taken aback by a fourth grader who said he'd like to try the steroids his sixth-grade brother uses. Previous education efforts have at times backfired; in Oregon, students who learned about the dangers of steroids were more likely to use them than those who didn't. Testing all high-school football players alone would cost $100 million and be nearly useless, since most teens know how to beat the tests with the use of "masking" drugs available underground.

At the forefront of education efforts are Charles Yesalis, professor of health policy at Pennsylvania State University and the nation's premier steroid expert, and Steve Courson, a former NFL star who used steroids. Both say that curbing steroid use requires nothing less than a revamping of American values. "We don't allow our kids to play games for fun anymore," says Yesalis. "We preach that God really does care who wins on Friday night, when we should be teaching our children to be satisfied to finish 27th, if that's their personal best."

Courson, in his recent book, "False Glory," tells of being introduced to steroids at age 18 by a college trainer, using steroids throughout his college and pro career, and developing an accelerated heartbeat during his heaviest cycle. He is currently awaiting a heart transplant. "In the NFL, I was nothing more than a highly paid, highly manipulated gladiator. I was spiritually bankrupt," says Courson, now a Pennsylvania high-school football coach. "I want kids to know they can be greater than gladiators, that they can use a sport to learn lessons about life and not let the sport use them."

Ultimately, to reach children, educators will have to crack the secretive steroid subculture. So inviting is the underground world that, according to one study, 1 in 100 users takes steroids primarily out of desire to belong to the tightknit group. Those who opt out are quickly ostracized. Bill, a 17-year-old junior from New England, says he was a wallflower with only a couple of friends before he got into steroids. Two and a half years and 16 cycles of steroid use made him part of the fellowship. But Bill vividly remembers one day last winter: It's the day his parents found a needle he forgot to discard. Since then, he hasn't seen much of his friends. "I had to switch gyms because they were all teasing me about shrinking up and pressuring me to use the stuff," he says. "I never see them now—we don't have anything to talk about anymore—but they're all betting I'll go back on it. Right now, the only way I know I'll stay off steroids is if I can find a guarantee that I'll reach 220 pounds without them. No, make that 230."

THOUGHT EVOKERS

- Explain the known physiological risks produced through the use of illegal steroids.

- What are the psychological conditions produced by steroid use?

- Are there any advantages in the use of steroids? Explain.

19 *Havoc* in the Hormones

Pollutants like dioxin and pesticides have upset the reproductive systems of alligators and gulls. Now, researchers theorize, the contaminants may be threatening humans.

By Jon R. Luoma

Louis Guillette, a University of Florida wildlife endocrinologist with salt-and-pepper hair and matching beard, is stalking an alligator, one of dozens with which we presently share a large cage.

This is not high adventure. The alligators are babies—the young of the year, a foot or so long. They scramble away from Guillette, scurrying among potted plants or diving and swimming furiously across the shallow pool in the center of the cage's concrete floor. They are fast and nimble, but Guillette soon has one in his hand, a male about 14 inches long. He grips it like a sort of huge, green hypodermic, nestling its torso in his palm, its head between two crooked fingers.

"There is no way in the world, just looking at this little alligator, that you could tell there was anything wrong with him," says Guillette, stroking the animal. The little reptile certainly seems robust, pumping its stubby legs in the air and squeaking the curious distress call—a quasi-electronic chirp—of a baby alligator in the clutches of predatory danger. But beyond its present predicament, something is indeed very wrong with it.

Guillette turns the alligator upside down and opens a pinkish genital structure on its abdomen. The oddity of this strikes me for a moment: that I have traveled all the way to Gainesville to look at an alligator's penis.

There isn't much to a baby alligator's penis; it is small, almost thread-

like. But although a layman wouldn't know it simply by looking, this one is smaller than it should be. And that's part of a larger story that perhaps should have us all sounding an alarm.

The upshot of the story is this: In the past decade, scientists like Guillette have been finding more and more evidence that molecules of some chemical pollutants—ranging from pesticides to polychlorinated biphenyls (PCBs) and from dioxins to additives in some plastics and industrial detergents—can behave like hormones in the bodies of both animals and humans. Mimicking natural hor-

---

"A chill went up my spine. It was the same pattern I'd seen in female alligators. I wondered where in the world they were getting estrogen."

---

mones (or blocking, amplifying, or otherwise disrupting them), these pollutants have already created biological chaos in the bodies of some exposed animals, like this alligator. And since such pollutants abound in ecosystems, as well as in animal and human tissues, some scientists worry that the problems seen so far represent only the tip of an iceberg.

Hormones are the body's chemical messengers. Secreted in infinitesimal amounts into the bloodstream, they can trigger dramatic changes in the body, like the alert, hair-raising, heart-

pounding sensation brought on by a surge of adrenaline, or the exquisite sequence of timing in the ovulation cycle. Hormones also exert enormous control over fetal development, enhancing the growth of skin, the nervous system, and some brain cells, and determining whether an embryo will develop the body of a male or a female.

A hormone molecule functions by docking onto a cell structure called a receptor, which has an affinity for the hormone's unique chemical design. Like a molecular key fitting into a specially designed keyhole, the hormone essentially flips a biochemical switch, which in turn can trigger a cascade of physical changes. But, says Guillette, "we now know for certain that some pesticides and other chemicals act like keys too—as fake messengers."

The alligator Guillette holds in the palm of his hand was born near Orlando on Lake Apopka. There, pesticide residues from agriculture as well as contaminants from a factory owned by the Tower Chemical Company continue to pollute the food chain even now, more than a decade after the plant was closed. As a bizarre consequence of that pollution, this alligator may not, in fact, be a male at all.

In terms of its hormonal chemistry, it seems to be closer to a female. Like other "males" born on the lake, its blood is high in estrogen and low in testosterone. As it matures, its penis could grow to as little as one-fourth the normal size; the animal is likely to be, like many of the Apopka alligators, "reproductively incompetent," as Guillette puts it.

It took Guillette and a group of colleagues years to piece together the puzzle of Lake Apopka's alligators. Since the mid-1980s they had known that something had caused the population to crash and that the alligators had strange anomalies. For instance, when he examined the ovaries of young female alligators, Guillette found that they were producing abnormal eggs, with multiple nuclei, and more than the usual one egg per egg follicle. A critical puzzle piece fell into place in 1992, when fellow endocrinologist Howard Bern visited Guillette's lab.

Bern, of the University of California, Berkeley, is a well-known expert on the now-banned drug DES, a synthetic estrogen that was prescribed for millions of women in the 1940s, '50s, and '60s; years later, the daughters of DES-exposed mothers began to have severe gynecological problems. During his visit, Bern showed Guillette and his students slides of the ovaries of laboratory mice that had been exposed to DES.

"A chill went up my spine," says Guillette. "It was the same pattern that I'd seen in my female alligators. But DES is a synthetic estrogen. I wondered where in the world my alligators were getting estrogen."

As Guillette accumulated data, the picture became clear: Pesticides (including dicofol—a close relative of DDT—and some DDT that had been produced as a contaminant during dicofol manufacture) had moved into the food chain of the Lake Apopka ecosystem. Studies by other scientists working with birds and laboratory rodents had shown that these kinds of pesticides, from a class of chemicals known as organochlorines, were not only tenacious once they entered animal tissues but also could mimic, amplify, or block natural sex hormones like estrogen and testosterone. Indeed, Guillette has been able to duplicate the resultant gender-bending effects in his laboratory by infusing clean alligator eggs with similar pesticides.

Guillette's discovery was far from isolated. If this were a problem with one pesticide and one species, the story of the alligators would be little more than a scientific oddity. But Theo Colborn, a zoologist and senior scientist with the World Wildlife Fund, says the phenomenon is all too common: Chemicals that disrupt the endocrine system now abound in ecosystems and, at least in the most polluted areas, are causing obvious problems. "We're seeing this same pattern of anomalies linked to endocrine disrupters across a whole suite of animals and humans around the world," she says.

In fact, from Europe's North Sea to the Great Lakes to the California coast, scientists have been piecing together remarkably similar puzzles, from seals with immune-system disorders to gulls with vestiges of both male and female gonads to fish that fail to mature sexually—all carrying high burdens of synthetic chemicals, most notably organochlorines.

Further, some scientists believe that the ubiquitous synthetic pollutants that virtually all of us now carry in our own fatty tissues could explain, for instance, why rates of breast and testicular cancer have soared, and why male sperm counts in the industrialized world have plummeted by a startling 50 percent since the dawn of the "chemical revolution" that began after World War II.

Theo Colborn has often served as the synthesizer who, beginning in the late 1980s, has helped bring together researchers who had made largely isolated discoveries about pieces of this puzzle. Yet for her, the discovery that the puzzle existed at all was an accident. Hired in 1987 by the World Wildlife Fund to assemble a report on Great Lakes environmental problems, Colborn began to pore over articles that highlighted problems observed in wildlife in the Great Lakes ecosystem: terns and double-crested cormorants born without eyes or with twisted, crossed beaks; salmon in polluted waters that failed to mature; gulls that showed no interest in nesting. She was intrigued by what she sensed was a pattern, but, she says, "I just couldn't sort out how all these pieces fit together."

Eventually, after she began plugging the data into a chart, a clearer pattern began to emerge. Virtually all the affected species were top predators, mostly fish eaters. Although data showed that they had relatively high levels of compounds like PCBs, dioxins, and a wide range of pesticides,

---

"People love to think that we're different from other animals. But at the cellular level, we are fundamentally the same."

---

most of the problems were not showing up as defects in adults. Rather, the adults were either failing to reproduce or producing offspring that failed to thrive. Eagles, for instance, that had recently migrated to the shores of industrially contaminated bays bore healthy chicks. Yet those that had fed on contaminated fish for years were either failing to reproduce or were bearing chicks with defects like beaks so mangled they were unable to eat.

Colborn also ran into studies by developmental psychologists Joseph and Sandra Jacobson at Wayne State University, in Michigan, suggesting that children born of mothers who had regularly eaten contaminated fish weighed less than average at birth, had smaller heads, and, as they matured, began to show decreases in short-term memory and attention span. Once again, the mothers seemed healthy.

Since the adults were fine, Colborn concluded that something was happening at the embryonic or fetal level, that somehow pollutants accumulated by the mother were affecting the offspring. But how? Just as Guillette, an endocrinologist, was led into the world of toxicology, more digging led Colborn, who was trained in wildlife toxicology, into the world of hormones.

Studies by Michael Fry, an avian toxicologist at the University of California, Davis, helped provide the tip-off. By the early 1980s, Fry had managed to replicate in his laboratory a bizarre anomaly he'd found in the 1970s among western gulls off the California coast. According to his research, male gulls in one DDT-contaminated area were ignoring breeding colonies, and females were nesting with females. (DDT use was restricted in the United States in 1972, but residues remain in food chains.) Dissecting some birds, he found that some males were functional hermaphrodites, with at least partially formed female sex organs. He was later able to reproduce this effect in his lab by infusing clean gull eggs with DDT or estrogens.

Colborn also found studies showing that laboratory rodents exposed to either natural hormones or a range of environmental toxicants, including dioxin, could literally have their sexual orientation altered. In one study by University of Wisconsin toxicologist Dick Peterson, male rats whose mothers had been injected with dioxin not only were disinclined to mate with females but moved into a femalelike mating position when in the presence of normal males. Although adults exposed to the pollutants seemed unharmed except at high dosages, the rats exposed in utero produced only half the usual amount of sperm. Even tiny doses caused effects in offspring.

In a 1985 study at the Environmental Protection Agency's Health Effects Research Laboratory in North Carolina, female hamsters exposed to Xearalenone, a common fungicide, developed what EPA scientist Earl Gray calls masculine behaviors, including an inclination to mount another female in estrus. The scientists believe that exposure to the poison may have altered the organization of cells in the brain that control sexual behavior.

In 1991 Colborn brought together many of the scientists studying this issue for a conference in Racine, Wisconsin. Among other things, participants agreed that pollutants can mimic or otherwise disrupt hormones, that "many wildlife populations are already affected" by endocrine-disrupting pollutants, and that because the same pollutants are accumulating in human cells, "a major research initiative on humans must be undertaken."

Colborn says, "This scares us all enough that we'd be delighted if someone could prove we're wrong." But she believes there is already enough evidence of risk to human health that federal regulatory agencies should insist that chemicals be tested for hormonelike effects before allowing them to be loosed upon the world.

Experimental studies can't be performed on humans. But the artificial estrogen DES may have served as a sad and unwitting experiment. Just as with wildlife and experimental animals exposed to toxins, the drug appears to have had little effect on most mothers who took it. But when their daughters reached maturity, many suffered from infertility, had malformed reproductive tracts, or were afflicted with an otherwise exceedingly rare cancer, clear-cell vaginal adenocarcinoma. DES sons showed an increased rate of problems such as undescended testicles, testicular cancer, and reduced sperm counts.

At a January 1994 conference on "environmental estrogens" in Washington, D.C., Danish researcher Niels Skakkebaek and British reproductive biologist Richard Sharpe reported that similar problems among men have been rising since midcentury, including a tripling of testicular-cancer rates in the United States and Britain and a plummeting rate of sperm production across the industrialized world.

"Why should we be concerned about what's happened to wildlife?" asks Robert Kavlock, who directs the EPA's developmental-toxicology division. "One word: *canaries* [a reference to the old coal miners' practice of using a canary as a monitor for poison gas]. We know there are problems with reproductive health among humans. It's hard to know what the cause is. But we know that if the problem is the environment, it's going to show up first in species with shorter life cycles than ours. We know there are hot spots for wildlife where we've seen these kinds of problems. The issue we've got to address is whether there are problems in the general environment."

Steven H. Safe, a Texas A&M chemist whose work is partially funded by the Chemical Manufacturers Association—a trade and lobbying group—is skeptical of the mounting research. In the April 1995 issue of the journal *Environmental Health Perspectives*, he flatly stated that "the suggestion that industrial estrogenic chemicals contribute to an increased incidence of breast cancer in women and male reproductive problems is not plausible." Although he agrees that reductions in such compounds in the Great Lakes "correlates with the improved reproductive success of highly susceptible fish-eating water birds," he points to naturally occurring hormonelike chemicals in some foods, like soy, as an estrogen mimic; further, he suggests that human intake of industrial compounds is too low to cause serious health problems.

However, Guillette and other researchers note that the human body has most likely evolved to cope with natural hormone mimics in foods, and that food-based compounds do not build up in fat cells, as do many of the synthetic industrial compounds. Guillette cautions, "Tying all of this into effects on the general human population at lower exposures is theoretical. But there are enough data to be concerned. There's no question that we have laboratory data showing that quite a few of these substances are endocrine disrupters. It's clear that in contaminated animals in the wild we have the same kinds of abnormalities we've produced in the lab. It's also true that in humans we're seeing increasing rates of these same kinds of problems. We're jumping ahead of the data if we try to say 'X causes Y.' But we aren't jumping ahead if we hypothesize and then support it with data. And it seems like the more data we collect, the more support we have. . . . So far, we have not found data that falsify the hypothesis.

"People love to think that we're different from other animals, and certainly different from insects. But at the cellular level, we are fundamentally the same. If we design a compound to be toxic to an insect cell, why does it surprise us when we find out that the same compound is toxic to a human cell? We've always thought the issue was mass—that these things could be toxic to an insect without having significant effects in a much larger human. But how big is an embryo?"

THOUGHT EVOKERS

- How have pollutants such as dioxin and pesticides upset the reproductive systems of alligators?

- How might human hormonal activity be affected by various toxic wastes?

- What are the functions of hormones? What is an endocrine gland? Explain.

VII

Ecology and Behavior

Ecology

> God made Heaven and Earth with all its murderous mountain fea-
> tures. With all its crawling and squirming creatures—equipped them
> with wings and musical screeches. And then he made man in his
> image—and man said, "I move the nominations be closed . . ."
>
> *Robert M. Chute*

From an ecological point of view, the human species is proving the most
destructive. Not only have we destroyed and poisoned other living systems, we have become our own victims as well. Produced as by-products of our technological skill, greed, and overpopulation, toxic wastes
are destroying the very components of our environment that we depend
upon for our existence. And the future is not promising: biocide applications, food additives, industrial and radioactive emissions, garbage, and
people pollution continue to accumulate. Examples of ecological devastation are reported in the article "Toxic Wasteland," which chronicles
occurrences in the former Soviet Union when economic growth was
worth any price. The price is proving enormous: 6 million acres of productive farmland lost to erosion; schools built on a radioactive waste
dump; 30 percent of all foods shown to contain hazardous pesticides;
and a radiation map, never before made public, pinpointing more than
130 nuclear explosions. The environmental destruction that took place
when radioactive material was released during the Chernobyl disaster
alone covered more the 50,000 square miles.

In the United States, the fitness of our land, air, wildlife, and aquatic
ecosystems—although receiving much political and media attention—
remains in poor condition. As addressed in the "27th Environmental
Quality Review" published by the National Wildlife Society, overall
environmental quality has not substantially improved during the past
27 years, and it continues to erode.

In effect, we are artificially selecting against the evolutionary survival of
many plant and animal species, and in so doing may be experimenting
with our own success as well; for it may come to pass that human animals are the ultimate endangered species.

Behavior

Ethologists (scientists who study behavior) are beginning to offer new
insights into the behavioral responses of humans and other living systems. Clues to the origin and treatment of depression and obsessive-
compulsive disorders are now suspected to be linked to the role of sero-

tonin neurons. Recent investigations reveal that clusters of serotonin neurons within the brain and the brainstem may partially account for the system's influence over such basic functions as the sleep cycle, sex drive, eating behaviors, body temperature, cardiovascular activity, and mood and aggression. Research based on the serotonin-neural linkage has produced a new generation of antidepressant drugs, including selective serotonin reuptake inhibitors such as Prozac, which result in the increase of the functional activity of present serotonin and may help reduce the symptoms of depression and obsessive-compulsive behavior.

Despite the intense current interest in human emotional conditions, not much research has been forthcoming regarding the nature and development of complex emotions. Ethologists now recognize that shame, and other self-conscious emotions such as guilt and pride, are involved in motivating human behavior. Congruent with these findings is the knowledge that failing (or succeeding) to live up to an internalized set of standards can produce extreme changes in human emotions.

20

From Conflict to Understanding

Forging a New Common Ground for Conservation in the 21st Century

By Jack Ward Thomas

It is my honor to share my thoughts with you—as a Professional Member of the Boone and Crockett Club and as Chief of the Forest Service—to renew the historic bond between the Forest Service and the Boone and Crockett Club.

I am proud to share the great conservation legacy of the Club, which has been a "keeper of the torch" for conservation. The Club, with its beginning in 1887, was among the Nation's first conservation organizations with early major credits in helping establish the Forest Reserves which were the forerunners of the National Forests, protection of Yellowstone National Park, establishment of wildlife protection laws, and advocacy for ethics of hunting and conservation. Later accomplishments included waterfowl protection, support for wildlife research and the wildlife refuge system, and the establishment of the "bible" for trophy records of North American big game. Many past members, including Theodore Roosevelt, George Bird Grinnell, Gifford Pinchot (the first Chief of the Forest Service), J.N. "Ding" Darling, and Aldo Leopold (who began his career with the Forest Service), are legends in American conservation. The Forest Service, the National Forests in its trust, and all Americans are the beneficiaries of this legacy. I thank you for that especially on behalf of the Forest Service.

I have been Forest Service Chief for nearly a year. The job is one that I neither sought nor wanted. But, I am honored to be Chief at this critical juncture in conservation history. Forest Service people have worked hard over this past year pushing in new directions that have evolved materially from the old. This direction is based on the very first statement of a land ethic by a land management agency. It reads, "promote the sustainability of ecosystems by ensuring their health, diversity, and productivity." And, on management and use of the National Forests and Grasslands tied to ecosystem principles. As you might expect, I've been dealing with the controversy associated with this change and the tension of folks who see themselves as potential losers, in the process of change.

Ecosystem management is an idea whose time has come. The concepts are old but the science, technology, philosophy, and sociopolitical situation now make it possible. Yet, I believe the English historian Henry Thomas Buckle had it just right when he said, "every new truth which has ever been propounded has, for a time, caused mischief; it has produced discomfort and often times unhappiness; sometimes disturbing social and religious arrangements, and sometimes merely by the disruption of old and cherished associations of thoughts . . . and if the truth is very great as well as very new, the harm is serious."

I was to spend 21 years in La Grande—professionally exciting and personally rewarding years. I intended to finish out my career there.

Perhaps some of you know me from my work as leader of the Interagency Scientific Committee to Address the Conservation of the Northern Spotted Owl. That committee developed a scientifically credible conservation strategy for the species. You may know of my continued involvement in northwest forest issues, including leadership of a series of ad hoc teams that performed scientific analyses of resource conditions and options for management. These assignments culminated in my chairmanship of the Forest Ecosystem Management Assessment Team (FEMAT) that produced an array of ten options for President Clinton's Management Plan for Federal Forests of the Pacific Northwest. This significant and difficult issue has been superficially portrayed and perceived as owls versus old growth, and jobs versus environment. This deflected understanding and appreciation of the real issues. That is, how do we use our natural resources so that all are sustainable, or how do we use the land without overspending our environmental capitals?

Aldo Leopold made the issue crystal clear—"The first precaution of intelligent tinkering is to save every cog and wheel." If, in our tinkering (using our natural resources) we carelessly throw away pieces, it is likely that—sooner or later—we could throw away the drive shaft. It is critical for us now to learn to husband our forest and range lands so that the beauty, integrity, and functions of those ecosystems are maintained century after century. This is no small task. And, it is, over the long term, the most critical task facing our kind as we increasingly exert dominion over the earth.

The Problem

This problem has broader dimensions. It is well to size up what is likely coming. There seems to be some things we should count on:

• The next two decades will be a time of significant world and national population and economic growth. International trade will increase significantly. World population seems likely to increase about 1 billion per decade. America's population will increase about 20 million per decade. This population growth coupled with economic growth will act in a synergistic fashion to up the pressure on natural resources.

•   The remainder of this decade will see concern with and addressing of environmental issues. . . . Some see the recent elections as a step back from environmental concerns. I think not. But, there will be demand for evolution of effectiveness. The "baby boom" generation will lead the nation from the mid 1990's to the mid-2010's. They are, and likely will remain in large part value driven, idealistic advocates. Pragmatic environmentalism will likely be a major part of their agenda. As leaders, they'll be strong activists and not quick to compromise.

The bottom line is that nearly all natural resources will become more scarce relative to demand. Simply, conservation must occur in the context of global and national economic growth structured to keep environ-

---

Ecosystem management is an idea whose time has come. The concepts are old but the science, technology, philosophy, and sociopolitical situation now make it possible.

---

mental transformation within safe limits. Since much of the slack will be out of the system, we will be dealing on the margins of both human and resource tolerances with ever-declining margins for error.

For all of us interested in conservation, and for Forest Service people in particular, these conditions intensify an old problem—how do we prepare for the likelihood of rapidly increasing human demands for wood, wildlife, minerals, clean water, grazing, and recreation over the next decades? How do we facilitate some inevitable level of economic growth, while

simultaneously becoming more environmentally and ecologically aware, as well as sensitive and responsive to public desire and demands ?

At present, the problem is already compelling and the signals from the public, politicians, law, and the courts are an increasingly volatile mixture. But this is also a time of "getting ready" or "getting fixed" as we say in the old country. The rest of the 1990's will likely be increasingly intense for those in the conservation business. Yet, the 1990's seem likely to be a time of relative calm before the storm of the early 21st century when it will become increasingly obvious that the time of reckoning has arrived.

The Solution

I am confident that we can begin dealing with effective solutions to these tough problems in our own backyard. One such solution was described in the theme of the National Forest System Centennial—which the Club so generously helped the Forest Service celebrate in 1991. Thank you for that vision and generosity. The theme, you may recall, was "Learning from the Past for a New Perspective of Conservation in the Future." The theme acknowledged and honored the wisdom of three of the Forest Service's and the Nation's conservation heroes from the past:

•   Aldo Leopold, who emphasized the importance of a land ethic and the significance of the application of ecology in land management and formatted the profession of wildlife management;

•   Gifford Pinchot, who emphasized public service, the creation of the National Forest system, and the wise sustainable use of natural resources; and,

•   Bob Marshall, who taught the importance of natural beauty and spiritual values of forests and was a seminal force in the establishment of a wilderness system.

When combined, the collective intelligence and philosophy of these great leaders can serve as the foundation for defining the levels and kinds of natural resource uses that are socially, economically, politically, and ecologically sustainable. We can visu-

alize the collective wisdom of Leopold, Pinchot, and Marshall to resolve the sustainability issue. This can be done in two major steps:

1. Development of broad-scale, ecosystem-based management strategies, that are routed in scientific knowledge and are ecologically, politically, and socially acceptable; and

2. Development of broad-based, regional forums that probe continuously for the common ground—the national consensus—for sustaining human uses and ecosystem form and function.

In January 1994, I signed, along with Jim Baca, then Director of the Bureau of Land Management, a letter directing that an ecosystem management framework and assessment of factors pertinent to management be developed for lands administered by the Forest Service and Bureau of Land Management east of the Cascade Crest in Washington and Oregon, and within the interior Columbia River Basin, which includes Idaho, western Montana, and portions of Wyoming, Utah, and Nevada. This unprecedented action was taken because management of the public's resources requires new direction based on ecosystem concepts within the context and other ownership and human uses within the larger basin. This "breaking new ground" will result in a new management strategy setting the framework for local plans and decisions. This approach will:

•   be based on concepts of ecosystem management

•   focus on maintaining or restoring the health of forest and aquatic ecosystems;

•   be scientifically sound;

•   consider the social, legal, economic, and political ramifications of proposed management approaches;

•   be a multi-agency effort, that fully involves the public in an open process; and,

•   be completed by the end of 1995.

This unprecedented effort has been in progress for nearly two years. I am confident this effort—the next evolutionary step to ecosystem management—will be an important step in the continuing process of evolving a set of standards that will provide reasonable estimates of safe environmental thresholds within which sustainable yields of goods and services can occur. I expect the adaptive management practices that result will provide constant midcourse corrections. I believe that if such an assessment and management strategy had been in place in the Pacific Northwest 10 years ago, the misnamed "spotted owl/old growth" issue could have been avoided. But, if we look back with the intent of learning and moving on, we can derive a better view of what the future can be. To look back to seek scapegoats is both unfair and debilitating. Rather, we need to focus on the fact that the future is ahead and we want to intelligently do what we can to shape that future.

At least we can hope the Columbia River Basin Project may help avoid development of similar divisive debilitating issues in that broader, precious area. Within that area lies my home to which I will return when my sojourn as Chief is complete. I have special

---

I believe there are grave risks to our precious public land ecosystems that should unify all who have been fighting over less serious issues.

---

concern for the place and its people. It and they are part of me. My wife of 38 years is buried there. I think that this new step in conservation is in the tradition of Leopold, Pinchot, and Marshall. In it the journey they began continues.

There has been a festering and debilitating national crisis over the use of public forest and range lands. The friction in this increasingly acrimonious struggle has chewed up a number of very good, talented, and dedicated professionals. That is one of the reasons that I am Chief. And, that same fate probably awaits me. That seems to be the way of things.

But naming a new Chief has not solved the crisis—for who was Chief

was not the root cause of the conflict. The deep polarization over how to manage—or not to manage—our natural resources issues continues and even deepens. This polarization creates the increasingly untenable situation within which public natural resources leaders and managers currently find themselves. And broadscale, ecosystem-based assessments and plans in and of themselves won't solve the problem either. Only when public understanding, acceptance, and support for a common ground for sustaining use and ecosystems comes about will those managers and leaders be able to move ahead with vigor and confidence. Recent polls have told us that the American people want to be consulted about how their lands are managed. But, after all, they expect the Forest Service to lead. We intend, and are, doing just that with increasing vigor.

An October 1994 national survey was conducted for *American Forests* to measure attitudes toward forest management and forest health issues. That poll indicated that the American public holds mixed views on forest management. They tended to believe that the Nation's forests are in reasonably good condition. About half were aware that the 1994 fire season was particularly severe. They tended to favor active forest management including salvage of burned trees and thinning to improve forest health. But they were closely divided on issues such as harvesting timber from Federal forests, allowing logging in burned-over areas to be delayed by real challenges, and use of controlled fire.

And the survey showed important regional differences. Respondents who lived in the West were more aware of current problems and more favorable toward timber harvesting and human management of forests than those in the South and Midwest.

Gifford Pinchot drove home the point to the Forest Service that the best-intentioned and best-reasoned conservation plans would fail without public support. He encouraged his foresters to work as hard with the public and the press (we would say

media today) as they did with resources. The Forest Service continues that tradition today, but development of consensus is elusive. Defining and standing on common ground is becoming increasingly difficult.

But, there is hope. The public and the professionals are beginning to weary of the extreme rhetoric from "the professional gladiators" whose interest is in the contest in the arena and not in consensus. From that weariness and the resulting impatience of those in political power there is, I believe, an atmosphere developing that can lead to that common ground, that consensus that has been missing for such a long and debilitating time. Almost surely the debate and the reexamination was healthy and was needed. But, now, a new need and a new day are coming. I can feel it. It is time. We must succeed in finding common ground in the public arena for sustaining both use and ecosystems. We, for the moment, seem to be without vision—and "without vision the people perish." But our vision is reforming anew and proceeding apace.

I believe that the solutions to the national crisis over use of public lands will be defined regionally and locally within the principles of ecosystem management. And, that implies that all who are concerned must be leaders in seeking for and defining the common ground for solutions to regional and local problems. Simply, the polemic debate over values (use versus preservation) must and will come to a conclusion. We must focus, as partners and as part of our democracy, on defining practical, regional, and local principles of management that can guide responsible shared resource use.

I noted the keen interest the intent in the Club's new strategic plan, "to participate in resolution of key natural resource conflicts through forums, symposia, workshops, and other appropriate means." I also am pleased with the December 5, 1990, Master Memorandum of Understanding between the Club and the Forest Service, declaring the groups as partners to accomplish mutually beneficial projects and activities. I trust that such includes a joint search for common ground in the management of public land. A marvelous opportunity is developing as we consider the ecosystems in the Idaho, Montana, Wyoming, Utah, and Nevada portions of the Columbia River Basin.

The Boone and Crockett Club has always squarely faced up to the key conservation challenges of the times. In 1887, the Club was one of a small handful of organizations seeking unity in conservation. Today, there are literally hundreds of conservation and environmental organizations. Many thrive on division and are threatened by unity. The conservation challenge of today is to bring about some consensus about natural resource management and a unity in resolving natural resources issues. I celebrate the Club as a partner in meeting this challenge. The Forest Service needs help in responsibly carrying out its mission of "Caring for the Land and Serving People." We have extended our hand to the Club—in friendship and in a plea for help—and you have grasped that hand.

I believe there are grave risks to our precious public land ecosystems that should unify all who have been fighting over less serious issues. There are suggestions heard in Washington to "sell off" or "give away" these lands that belong to us all. Such talk strikes at my heart and tears at my vitals. I was raised in a state with no public lands to speak of. I stepped into a National Forest for the first time when I was in my 30's. And, my life and view of land have never been the same since. This was *my* land. I was part owner and it stretched out before me and I could go where I liked without begging permission. It was my land and my children's land, and their children's children's land. No other nation has such a heritage. It is a heritage that is worth constant vigilance—and even a fight now and again. Surely, there is common ground there in those lights. In saying that I remain cognizant of the role that the Club played in putting the original Forest Reserves in place and the role that a Club president named Gifford Pinchot played in putting the Forest Service in place. It is a proud heritage—for the Boone and Crockett Club and for the Forest Service.

I pledge the full power, knowledge, and influence of the Forest Service to work with the Club to explore ways of developing a sorely needed public unity based on the principles of land ethics, shared use, and ecosystem sustenance to reduce these risks. With this focus we can build on the dreams of the Club's founders to the lasting benefit of our children and the generations who follow. That, my friends, is a worthy challenge and a noble pursuit.

My Texan grandfather had a saying he applied to those few that he held in highest esteem. I too use the expression—and, as he did, rarely. I say it to the members of the Boone and Crockett Club, "you will do to ride the river with."

THOUGHT EVOKERS

- Describe how we can develop a mainstream effort to maintain ecosystem diversity from conflict regarding environmental conditions.

- What programs need to be established in order to correct past environmental mistakes? Explain.

21

The Alarming Language of
Pollution

Evidence is mounting that false signals from synthetic chemicals are harming reproduction, immunity, and behavior in wildlife. Are humans also at risk?

By Daniel Glick

On California's Channel Islands in the mid-1970s, an ecologist found an abnormally high ratio of female gulls to male gulls. In Florida in the early 1990s, a team of endocrinologists discovered abnormally small penises in alligators near a former Superfund site. And in Great Britain, biochemists have noticed in the last few years that something in wastewater effluent appears to be creating hermaphroditic fish.

Sound like bizarre episodes of *Wild Kingdom*? Actually, these observations are all clues to a far-flung scientific sleuthing saga. Over the last few years, experts from a dozen disciplines have been piecing together field and laboratory evidence that environmental pollutants may be doing far more damage to wildlife and humans than previously suspected, in ways no one had imagined possible. For starters, by sending various false signals to endocrine (or hormonal) systems in the body, pollutants could be harming vertebrate reproduction worldwide. All of this evidence could comprise one of the most alarming messages wildlife has ever sent our way. "If we don't believe that animals in the wild are sentinels for us humans, we're burying our heads in the sand," says Linda Birnbaum, director of the environmental toxics division of the Environmental Protection Agency (EPA).

Endocrine-disrupting chemicals are associated with problems ranging from developmental deficiencies in children, to smaller penises in pubescent boys, to infertility. "Every day, I get more concerned," says John McLachlan, chief of the laboratory of reproductive and developmental toxicology at the National Institute of Environmental Health and Science (NIEHS).

Implicated are huge numbers of products—including some pesticides, industrial solvents, adhesives, and plastics. A very few, such as PCBs and the pesticide DDT, have been banned or are more heavily regulated in this country than in the past—though they persist in the environment. But thousands have never been regulated. Much of the stuff is deposited worldwide by the atmosphere and has been found in both the Arctic and Antarctica.

Until the last few years, the biggest question for regulators has been: Does a given chemical cause cancer, and if so, at what exposure level? (And very few chemicals have even been tested for carcinogenicity.) Now some researchers are also asking: Does a chemical harm reproduction, immunity, behavior, or growth?

Also, regulators have long assumed each chemical to be innocent until proven guilty. But researchers are growing increasingly concerned at evidence that related chemicals may be able to harm the body in similar ways. For example, DDT and dioxins (often commonly referred to in the singular) are members of a group of similar chemicals called organochlorines. They are not to be confused with the chlorine we safely use to disinfect swimming-pool water and bleach our clothes. While DDT is a deliberate product, dioxins are unwanted byproducts of industrial high-temperature use of chlorine.

A 1994 National Wildlife Federation report, *Fertility on the Brink: The Legacy of the Chemical Age*, concluded that there is enough evidence to warrant phaseouts, at the very least, of certain chemicals released into the environment. The list includes dioxins, some pesticides, and hexachlorobenzene. Federation counsel Elise Hoerath argues that the problem has become "a significant public health threat."

Others warn that hormonal activity is so complicated and poorly understood that costly action to ban certain chemicals is uncalled for until we know more. "As a citizen, I would like to see some of these chemicals banned," says Carlos Sonneschein,

> "If we don't believe that animals in the wild are sentinels for us humans, we're burying our heads in the sand."

professor of cellular biology at Tufts University School of Medicine. "As a scientist, I would like to have more data."

Still, the data have been steadily adding up, thanks largely to the work of zoologist Theo Colborn, a senior scientist at the World Wildlife Fund and director of its wildlife and contaminants program. In late 1987, Colborn began sifting through studies of declining wildlife populations in the Great Lakes region. On the left side of a piece of paper, she listed species with steep population drops: bald eagle, Forster's tern, double crested cormorant, mink, and river otter, among others. On the right, she listed their health problems, including organ damage, eggshell thinning, hormonal changes, and low birth survival rates.

Each of the animals depended on a fish diet. Fish in the notoriously polluted Great Lakes were known to contain high concentrations of various synthetic chemicals, especially in fatty tissue, and Colborn wondered if the pollutants were causing the disorders. Were toxics tinkering with the immunity, behavior, or growth of fish eaters? Colborn began searching the scientific literature. "I was really concerned," she recalls. "It was very obvious that these chemicals were developmental toxicants." Yet for the most part, testing had only looked for cancer. "We've been blinded," she says. "We never tested for developmental effects."

Even so, some studies did find those effects. Researchers had found in the mid-1970s that exposure to DDT seemed to be correlated with an abnormally low number of males in a California gull population. In the late 1970s, toxicologist Michael Fry of the University of California at Davis was able to cause "feminization" of male gull embryos (they developed abnormal testes containing ovarian tissue) in his lab by injecting uncontaminated eggs with DDT.

Many years later, in the early 1990s, University of Florida comparative endocrinologist Louis Guillette started finding similar problems in alligators at Florida's Lake Apopka. The area was a former Superfund site that had

---

"Very, very low levels of contaminants can have an effect on developing embryos."

---

been contaminated in 1980 with the chemical dicofol, an organochlorine that also contained some DDT. The lake also contained a mix of agricultural chemicals from farm runoff.

Working with colleague Timothy Gross and other researchers, Guillette found that alligator eggs were barely hatching, teenage males had abnormally small penises, and the level of the male hormone testosterone was far below normal. Later, Guillette conferred with a researcher who had produced remarkably similar results in lab rats by exposing them to a com-

pound similar to DDE, a breakdown product of DDT. "Oh my God," Guillette said after seeing the data. "I think we have a major problem here."

As Colborn compiled evidence from wildlife biologists, toxicologists, and the medical literature, she realized that other scientists were asking some of the same questions. So, in 1991, she helped bring a group of them together to compare notes for the first time. After another meeting last year in Washington, D.C., 23 wildlife biologists agreed that "populations of many long-lived species are declining. . . . Some of these declines are related to exposure to man-made chemicals and their effects on the development of embryos."

Their reasoning is based on the knowledge that sex differentiation is determined by tiny amounts of male and female hormones interacting in the developing fetus. Contrary to what we've all been taught in introductory biology classes, animals do not exhibit male or female traits simply because they possess or lack a Y chromosome. If a hormone impostor shows up during fetal development, sexual function can go akimbo. "Very, very low levels of contaminants can have an effect on developing embryos," says the University of Florida's Guillette. "A dose that wouldn't bother an adult can be catastrophic to an embryo."

Soon after Fry's discovery that DDT injections could "feminize" gull eggs, biologist David Crews of the University of Texas discovered in 1984 that he could control the gender of slider turtles with minute quantities of the female hormone estradiol. For many turtles the temperature of the eggs' environment determines gender. Heat produces a female; cold yields a male. But in the lab, Crews could coax embryos incubating at a male-producing temperature to become female with just a drop of estradiol on the eggs.

Estradiol is an estrogen, and Crews' study fits a scary pattern. A number of synthetic substances are so-called "environmental estrogens," acting like the hormone Crews used to bend the turtles gender. In recent work, he and colleagues have found they can create sexually mixed-up turtles with "cock-

tail" mixtures of certain PCB compounds. Some of the turtles have testes and oviducts. Others have ovaries but no oviduct. Most alarming, these effects occurred at extremely low doses. Somehow, the combination of several PCBs is far more disrupting than one PCB compound alone.

Of course, not all estrogens are bad; when they occur naturally, they play critical roles in the body. Deliberate therapeutic doses even help women through and beyond menopause, in part by protecting bone density and cardiac health. Environmental estrogens, however, are a different story. NIEHS researcher McLachlan, who calls estrogen the "Earth Mother of hormones," has shown that certain chemicals can bind to or block estrogen receptors, which may in turn cause developmental deviations.

Think of the estrogen receptor as a lock on a cell, and natural estrogen as a perfect key. Scientists believe that literally hundreds of compounds have a chemical structure that also fits the lock—and which could produce similar responses. But then, these chemicals may "fit" into estrogen receptors without producing the cascade of cellular events that follow exposure to actual estrogen—and no harm may be done. Still even if that's so, when the imposter key is in the lock, the real key may not be able to enter.

Since the number of chemicals that fit into the estrogen lock, or receptor, are so numerous, no one can clarify all the effects of these multiple exposures. "If there are so many estrogens out there, how can anybody figure out which one is doing what?" asks Thomas Goldsworthy of the Chemical Industries Institute of Toxicology. "Some of the mechanisms aren't clear yet."

Some of the effects, however, are becoming clearer. Toxicologists Earl Gray and Bill Kelce of the EPA reported last year that the common fungicide vinclozolin, used on many fruits and vegetables, can block receptors for the male hormone androgen and cause sexual damage in male rats. At certain doses, rats exposed to vinclozolin do not develop normal male traits even though they do produce testosterone. At high exposures, male rats develop severely abnormal genitalia. Gray thinks fruit treated with the fungicide does not contain enough residue to harm humans, but he is looking into the question. And he is sure of one thing: "There are clearly other environmental anti-androgens we haven't discovered yet," he says.

The findings of field work like Guillette's and laboratory analysis like Gray's have been bolstered by studies of inadvertent human exposures to endocrine-disrupting compounds. In 1979, women in Taiwan who ate rice oil contaminated with polychlorinated biphenyls (PCBs) and polychlorinated dibenzofurans (PCDFs) offered an ideal if tragic laboratory to track long-term effects in humans. Researchers have followed 118 children of the women and an identically sized control group. Members of the exposed group have suffered developmental delays, growth retardation, and slightly lower IQs. Many of the boys, who are now reaching puberty, have abnormally small penises.

Between the 1940s and 1970s, diesthylstilbestrol, or DES, was given to an estimated two million to six million women during pregnancy to help prevent miscarriage. In children of DES mothers, the drug caused a range of developmental and health problems, some of which only surfaced in the process of creating the next generation. Among males, researchers have noted abnormalities in scrotums, an unusually high prevalence of undescended testicles, and decreased sperm counts. Among DES daughters, clinical problems include organ dysfunction, reduction in fertility, immune-system disorders, and other difficulties.

The DES example leads to an alarming hypothesis: If some endocrine-disrupting pollutants act like DES, which had effects long after birth, perhaps we won't see the consequences until exposed offspring themselves begin trying to have kids. And that raises the question: What actual harm to humans have scientists found from exposure to the sea of chemicals released into the environment over the past 50 years?

Enter Niels Skakkebaek, a Danish researcher in Copenhagen. In 1991, he published a meta-analysis of many smaller studies of global human sperm counts over the past half century and found that the counts declined by half between 1940 and 1990. Other, more recent European studies sought to disprove Skakkebaek's results, but ended up corroborating them. If sperm counts have indeed dropped, one clue to the reason may come from lab tests in which estrogen-mimicking compounds have affected the Sertoli cell, which is related to sperm production.

Research has also implicated environmental toxics in the rise of endometriosis, testicular cancer, and possibly other cancers as well in recent decades. In one study that went on for 15 years, 79 percent of a rhesus monkey colony exposed to dioxin developed endometriosis (the development of endometrial tissue in females in places it is not normally present). Dioxin is not thought to imitate estrogen, but is clearly an endocrine dis-

"As a citizen, I would like to see some of these chemicals banned. As a scientist, I would like to have more data."

ruptor in at least some animals. In the monkeys, the endometriosis increased in severity in proportion to the amount of dioxin exposure.

What should the rest of society do while the researchers compare notes? "The tough call isn't for the scientists now," says Devra Lee Davis, a top scientific advisor at the U.S. Department of Health and Human Services. "It's for the regulators." There are signs that the federal government is beginning to pay heed. In the EPA draft dioxin reassessment report, now under review, dioxin is characterized as a potent toxic "producing a wide range of effects at very low levels when compared to other environmental contaminants."

The International Joint Commission, a bilateral organization that advises on environmental issues along the U.S.-Canada border, has repeatedly called for virtual elimination of toxic substances in the Great Lakes region. And a little-noticed amendment to the Clean Water Act proposed by the Clinton administration (the reauthorization died in the last Congress) would have required regulators to look at "impairments to reproductive, endocrine and immune systems as a result of water pollution." Even skeptic Goldsworthy of the Chemical Industries Institute of Toxicology says, "We are changing our environment. There's no question about that."

The World Wildlife Fund's Colborn says she welcomes scientific skepticism and even has days when she hopes she is imagining the whole thing. "We admit there are weaknesses, because we are never going to be able to show simple cause-and-effect relationships," she says of the complicated theory. Still, she adds, "The research has reached a point where you can't ignore it any more, and new evidence is coming in every week." For visitors to her Washington, D.C., office, Colborn lets a pesticide manufacturer have the last word: On the wall hangs a 1950s label from a one-pound package of a substance called DuraDust, 50 percent of which was pure DDT. The label promises, "Its killing power endures."

THOUGHT EVOKERS

- How can toxic wastes affect hormonal balances within various species of animals?

- Why has the use of DDT been outlawed in the United States?

- What is estradiol, and how does it affect the reproductive capability of sexually reproducing mammals? Explain.

22

TOXIC WASTELAND

*In the former Soviet Union, economic growth was worth any price.
The price is enormous.*

By Douglas Stanglin, with Victoria Pope, Robin Knight, Peter Green,
Chrystia Freeland, and Julie Corwin

In satellite photos of the Eurasian landmass at night, the brightest pools of light do not emanate from London, Paris, or Rome. The largest glow, covering hundreds of thousands of acres and dwarfing every other light source from the Atlantic to the Pacific, can be found in the northern wilderness of Siberia, near the Arctic Circle. It comes from thousands of gas flares that burn day and night in the Tyumen oil fields, sending clouds of black smoke rolling across the Siberian forest. During the past two decades, the steady plume of noxious sulfur dioxide has helped to ruin more than 1,500 square miles of timber, an area that is half again as large as Rhode Island.

Siberia's acid rains are just one more environmental catastrophe in a land where man has run roughshod over nature and is now facing the deadly consequences. The former U.S.S.R. had no monopoly on pollution and environmental neglect, as residents of Minamata, Mexico City, and Love Canal can testify. But Soviet communism's unchecked power and its obsessions with heavy industry, economic growth, national security, and secrecy all combined to produce an environmental catastrophe of unrivaled proportions.

"When historians finally conduct an autopsy on Soviet communism, they may reach the verdict of death by ecocide," write Murray Feshbach, a Soviet expert at Georgetown University, and Alfred Friendly Jr. in their new book, "Ecocide in the U.S.S.R." (Basic Books, $24). "No other great industrial civilization so systematically and so long poisoned its air, land, water, and people. None so loudly proclaiming its efforts to improve public health and protect nature so degraded both. And no advanced society faced such a bleak political and economic reckoning with so few resources to invest toward recovery."

In the name of progress. Communism has left the 290 million people of the former Soviet Union to breathe poisoned air, eat poisoned food, drink poisoned water, and, all too often, to bury their frail, poisoned children without knowing what killed them. Even now, as the Russians and the other peoples of the former U.S.S.R. discover what was done to them in the name of socialist progress, there is little they can do to reverse the calamity: Communism also has left Russia and the other republics too poor to rebuild their economies and repair the ecological damage at the same time, too disorganized to mount a collective war on pollution, and sometimes too cynical even to try. Even when the energy and the resources needed to attack this ecological disaster do materialize, the damage is so widespread that cleaning it up will take decades. Among the horrors:

• Some 70 million out of 190 million Russians and others living in 103 cities breathe air that is polluted with at least five times the allowed limit of dangerous chemicals.

• A radiation map, which has never been released to the public but which was made available to *U.S. News*, pinpoints more than 130 nuclear explosions, mostly in European Russia. They were conducted for geophysical investigations, to create underground pressure in oil and gas fields, or simply to move earth for building dams. No one knows how much they have contaminated the land, water, people, and wildlife, but the damage is almost certainly enormous. Red triangles on the map mark spots off the two large islands of Novaya Zemlya where nuclear reactors and other radioactive waste were dumped into the sea. Tapping one location, Alexei Yablokov, science adviser to Russian President Boris Yeltsin, says a nuclear submarine sank there 10 years ago, its reactor now all but forgotten. "Out of sight, out of mind," he says with disgust.

• Some 920,000 barrels of oil—roughly 1 out of every 10 barrels produced—are spilled every day in Russia, claims Yablokov. That is nearly the equivalent of one Exxon Valdez spill every six hours. To speed up construction of oil pipelines, builders were permitted to install cutoff valves every 30 miles instead of every 3, so a break dumps up to 30 miles worth of oil onto the ground. One pool of spilled oil in Siberia is 6 feet deep, 4 miles wide, and 7 miles long.

• According to Yablokov, the Siberian forests that absorb much of the world's carbon dioxide are disappearing at a rate of 5 million acres a year, posing a bigger threat to the world environment than the destruction of the Brazilian rain forests. Most of the damage is caused by pollution and by indiscriminate clear-cutting, mostly by foreign companies in soil that can't tolerate such practices.

• Officials in Ukraine have buried 400 tons of beef contaminated by radiation from the Chernobyl nuclear accident. An additional 920 tons will be buried in June.

A confidential report prepared by the Russian (formerly Soviet) Environment Ministry for presenta-

Thirty percent of all foods contain hazardous pesticides.

tion at the Earth Summit in Rio de Janeiro this summer blames the country's unparalleled ecological disaster primarily on a policy of forced industrialization dating back to the 1920s. The report, a copy of which was obtained by *U.S. News,* notes the "frenetic pace" that accompanied the relocation of plants and equipment to the Urals and Siberia during World War II and their rapid return to European Russia after the war. This, the report says, created a "growth-at-any-cost mentality."

The communist state's unchallenged power also was reflected in its obsession with gigantism and in its ability to twist science into a tool of politics. The late Soviet President Leonid Brezhnev planned to reverse the flow of the Irtysh River, which flows north, in order to irrigate parts

Two kindergartens in Estonia were built on a radioactive waste dump.

of arid Central Asia for rice and corn growing. But to redirect 6.6 trillion gallons of water each year would have required building a 1,500-mile canal.

Critics warned that the project would alter world weather patterns, but Soviet officials gave up only after spending billions of rubles on the plan. "Soviet science became a kind of sorcerer's apprentice," write Feshbach and Friendly.

Unexplained anthrax. Not surprisingly in a nation obsessed with national security and secrecy, another culprit was the military-industrial complex, which the Environment Ministry's report says "has operated outside any environmental controls." In 1979, some 60 people died in a mysterious outbreak of anthrax near a defense institute in Sverdlovsk (now renamed Ekaterinburg). After years of Soviet denials of any link with defense matters, the Presidium of the Supreme Soviet voted in late March to compensate the victims of the incident and conceded that it was linked to "military activity."

At the same time, the report says, communism's reliance on central planning and all-powerful monopolies produced an "administrative mind-set" that created huge industrial complexes that overtaxed local environments. The report says the emphasis on production over efficiency has led to some 20 percent of all metal production being dumped—unused—into landfills. Nor did Soviet industries, shielded from competition, feel any need to improve efficiency or switch to cleaner, more modern technology.

Worse, it became virtually impossible to shut down even the worst offenders, because doing so could wipe out virtually an entire industry.

In Estonia, for example, the Kohtla-Jarve chemical plant, a major polluter, squeezes 2.2 million barrels of oil a year from shale and provides 90 percent of the energy for the newly independent country. Environment Minister Tanis Kaasik says flatly that it is "impossible" to shut down production.

Terrible secrets. A pervasive secret police force, meanwhile, ensured that the people seldom found out about the horrors visited on them in the name of progress and that, if they did, they were powerless to stop them. It took Soviet officials more than 30 years to admit that an explosion had occurred at a nuclear storage site near Chelyabinsk in 1957. The blast sent some 80 tons of radioactive waste into the air and forced the evacuation of more than 10,000 people. Even with glasnost, a cult of silence within the bureaucracy continues to suppress information on radiation leaks and other hazards. Indeed, the No. 1 environmental problem remains "lack of information," says former Environment Minister Nikolai Vorontsov.

Even now, with the fall of the Communist Party and the rise of more-democratic leaders, there is no assurance that communism's mess will get cleaned up. Its dual legacy of poverty and environmental degradation has left the new political leaders to face rising demands for jobs and consumer goods, growing consternation about the costs of pollution, and too few resources to attack either problem, let alone both at once.

Although 270 malfunctions were recorded at nuclear facilities last year,

---

A Swath of Destruction

The environmental destruction wrought by the Soviet state stretches across thousands of miles—from radioactive soil in Ukraine to poisoned fish in the Volga to disappearing forests in Siberia.

• **Radioactive contamination**

More than 50,000 square miles were contaminated by radioactive material released in the Chernobyl disaster alone.

• **Aging nuclear reactors**

Ten Chernobyl-type reactors and six other poorly designed power plants threaten to leak radiation.

• **Air pollution**

Occasionally, 103 cities exceed air pollution limits by a factor of 10 or more.

• **Highly polluted waters**

Industrial wastes and untreated sewage have turned many lakes, rivers, and seas into a deadly mix of toxic chemicals.

• **Forest damage**

Inefficient processing and harvesting methods have led to deforestation and topsoil erosion. Five million acres are disappearing in Siberia every year.

• **Soil damage**

Severe erosion and indiscriminate use of pesticides have depleted or poisoned millions of acres of farmland.

economic pressure will make it difficult to shut down aging Soviet nuclear power plants. In March, radioactive iodine escaped from a Chernobyl-style plant near St. Petersburg, prompting calls from German officials for a shutdown of the most vulnerable reactors. Yeltsin adviser Yablokov warns that "every nuclear power station is in no-good condition, a lot of leaks." In the short term, Russia has little choice but to stick with nuclear power, which provides 60 percent of the electricity in some regions.

Environmental consciousness has permeated only a small fraction of society, and rousing the rest will require breaking the vicious circle of social fatalism. "We haven't got any ecological culture," says Dalia Zukiene, a Lithuanian official. Russian aerosols still contain chlorofluorocarbons, though Russia has now banned them, but if a Russian is lucky enough to find a deodorant or mosquito repellent, he will grab it—regardless of the consequences to the ozone layer. "We still bear the stamp of *Homo sovieticus*—we're not interested in the world around us, only in our own business," says Zukiene. Adds Alla Pozhidayeva, an environmental writer in Tyumen, in the oil fields of western Siberia: "Sausage is in the first place in people's minds."

Despite the mounting toll, the environmental activists who rushed to the barricades in the early days of glasnost have largely disappeared. When the Social Ecological Union recently tried to update its list of environmental groups, it found that more than half of them had disbanded in the past year. "If people go to a meeting at all, it isn't for the sake of ecology," says Vladimir Loginov, an editor of Tyumen Vedomosti, a newspaper in the Tyumen oil region. "They have to eat."

In fact, the crisis of leadership afflicting much of the former Soviet Union poses a whole new set of threats to the environment. The loosening of political control from Moscow already has turned the provinces—especially Siberia—into the Wild West. Local authorities, par-

ticularly in the Far East, have extended vast timber-cutting rights to foreign companies, especially Japanese and South Korean, without either imposing strict controls on their methods or requiring reforestation. "The economic chaos here presents enormous opportunities for local administration, without any government control, to cut forest, to sell it abroad, and to receive some clothes, cars, video equipment," says Yeltsin adviser

---

Scientists recently found 11 more areas poisoned by Chernobyl.

Yablokov. If you visit the Far East forest enterprises, you will be surprised how many Japanese cars you will find."

The breakup of the Soviet Union is adding to the tensions. Despite Chernobyl, Ukraine, facing an energy crisis as the price of the oil it imports from other regions rises to world levels, is quietly contemplating building new nuclear power plants. But a stepped-up Ukrainian nuclear power program would create its own problems: Krasnoyarsk, the traditional dumping ground in Russia for nuclear waste, is refusing to accept Ukraine's spent reactor fuel because Ukraine is demanding hard currency for its sugar and vegetable oil.

In the mountainous Altai region of Russia, which recently declared itself autonomous and elected its own parliament, newly elected officials are trying to revive a controversial hydroelectric project on the Katun River. Victor Danilov-Danilyan, the Russian minister of ecology and natural resources, says local officials in Altai, many of whom are former Communist Party leaders, are now trying to cast the battle over the project as a nationalist issue. He says local authorities have deliberately ignored the danger of increased toxic wastes in the water and intentionally underestimated both how much the project will cost and how long it will take to build.

"They're just deceiving people," Danilov-Danilyan charges. "They just want to grab as much as they can while they're in power, to build dachas for themselves."

Still, there are some glimmers of progress, including the recent creation of three new national parks in Russia. In February, President Yeltsin signed a new environmental law that empowers local officials or even individuals to sue an offending enterprise and demand its immediate closure. It also holds polluters, not some distant ministry, responsible for their actions. The new law further permits aggrieved parties to sue for damages, not just fines. The environmental ministry's report notes that over the years, "few ministries, if any, chose to clean up their act and didn't go beyond paying lip service to the need to protect the environment." In most cases, polluters got off with small fines or escaped punishment altogether by passing the buck to government ministries .

But Vladislav Petrov, a law professor at Moscow State University and the main author of the new legislation, says that if it is strictly enforced, the law would shut down 80 percent of the country's factories overnight. In the sooty steel town of Magnitogorsk, in the Urals, an independent radio journalist says he will try to force the Lenin Steel Mill, which employs 64,000 people, to close. He doubts he will succeed

Growth industry. Moreover, while the new, 10,000-word statute has teeth, only a handful of lawyers, and even fewer judges, are familiar with environmental law. Petrov says the courts are ill-equipped to handle claims from individuals and would be overwhelmed if people tried to collect damages from polluters. "In order for this article of the law to be effective, the whole court system should be changed," he says.

Still, environmentalism is a growth industry in the former Soviet Union. Many scientists in fields such as nuclear physics hope to recast themselves as ecologists. Mindful that the Russian government does not have the funds for large projects, they are looking for foreign partners to join them in cleanup projects. So far, most Western groups have offered advice but not much money.

Some Western input may be necessary, however, to prevent the environmental effort from succumbing to its own form of gigantism. One Central Asian academic's plan for saving the Aral Sea, for example, calls for building a 270-mile canal from the Caspian Sea to divert water into the depleted Aral. But because the Caspian Sea is lower than the Aral, the water would have to be pumped into the canal, and that would require considerable electricity. The proposed solution: build a network of solar power stations.

The spreading ecological disaster may yet force change on an impoverished and cynical people. "We have a Russian saying: 'The worse, the better,'" says Yablokov. "This situation has now become so obvious for all people that I feel that a lot of decision makers began to turn their minds in this direction." The Stalinist idea, he says, was to build socialism at any cost because afterward there would be no more problems. "It was an unhealthy ideology," he says. "Now I feel that my people are coming to understand the depths of this tragedy."

THOUGHT EVOKERS

- Describe the conditions that have contributed to the former Soviet Union's nuclear waste problem.

- Describe and explain the environmental legacy that the political leaders of the former Soviet Union have left to future generations.

23

27th Environmental Quality Review
A Year of Gridlock

By almost any measure, 1994 was a disappointing year for those Americans who looked to federal lawmakers for strong leadership on environmental and conservation matters. Despite President Clinton's pledge last Earth Day to "reinvent the way we protect the environment," the administration was repeatedly stymied by a U.S. Congress stuck in legislative gridlock over a long list of environmental matters. Several important measures remained stalled at year's end, including bills designed to give Cabinet status to the Environmental Protection Agency and to create a Biological Survey in the Interior Department. The lawmakers also failed to protect public lands from mining and grazing abuses and to pass revised versions of the Clean Water and Endangered Species acts.

"Congress is not doing its job of protecting the health and natural resources of the American public," observed National Wildlife Federation President Jay D. Hair as 1994 came to a close. "All too often, lawmakers are letting the nation's environmental agenda become subordinate to the needs of special interests."

A major factor in Congress' environmental failures in 1994 was opposition to strong protections by a so-called property-rights movement, backed by industry and big-time agriculture. Its goal: to weaken laws designed to protect natural resources, the environment, and public health. One of the favorite themes of the movement was that property owners should be compensated for any decrease in value of their property due to government regulation. "A tsunami of anger is rolling across the land," wrote *The Wall Street Journal* last year.

Tsunamis are caused by earthquakes, but the epicenter of this temblor was not easy to find. According to a nationwide poll conducted in 1994 for Times-Mirror Magazines, 76 percent of Americans thought regulations to prevent water pollution had not gone far enough, 51 percent wanted a strengthened Endangered Species Act, and 82 percent favored reform of the federal Mining Act. Public anger over federal protection of the environment could be found only among the groups that profit from exploiting natural resources. "They succeeded in creating the illusion of a rebellion where none exists," read an editorial in *The Atlanta Constitution* last August.

On the following pages, *National Wildlife*'s twenty-seventh annual Environmental Quality Review examines some of the past year's key events.

Wildlife

New studies show the advantages of protecting biodiversity, but the U.S. Congress fails to take heed.
After decades of decline, the future of America's national symbol, the bald eagle, looks considerably brighter these days. As a result of years of protection and recovery efforts under the Endangered Species Act and other laws, more than 4,000 nesting pairs now survive in the lower 48 states— 10 times more than existed three decades ago.

Last Fourth of July, the U.S. Fish and Wildlife Service (FWS) formally proposed that the eagle's status be downgraded from endangered to threatened everywhere in the lower 48 states except the Southwest. "The Endangered Species Act was an important tool in rebuilding eagle populations," said NWF researcher Steve Torbit.

A month earlier, federal officials had removed the California gray whale from the endangered list—the first time a marine creature had recovered strongly enough to be removed. Conservationists viewed such wildlife recoveries, in the words of Environmental Defense Fund attorney Michael Bean, as "important milestones" for the species and "also for the Endangered Species Act."

The effectiveness of the law was further documented by a FWS report last summer, which found that nearly 40 percent of the plants and animals protected under the act are now stable or recovering. "These statistics tell us we are making significant progress toward restoring endangered and threatened species," FWS director Mollie Beattie told Congress.

Conservationists hoped such progress would help galvanize support for the reauthorization by Congress of the Endangered Species Act, which was due in 1993 but was again postponed last year. After congressional opponents of the law tried to introduce provisions to weaken its effectiveness by requiring the government to compensate landowners for presumed damages caused by complying with the act, the reauthorization process came to a halt.

As the so-called property rights issue unfolded last summer, Interior Department Secretary Bruce Babbitt pointed out that of 118,000 development projects reviewed in the last two decades for conflicts with the law, only 33 had been rejected. He also announced new guidelines for administering the act, including measures to better consider the economic costs of protecting species.

Meanwhile, scientists continued in 1994 to uncover concrete evidence to support measures for safeguarding rare species and maintaining biodiversity. Cornell University researcher Thomas Eisner, for example, discovered in a rare and disappearing mint-like plant, *Dicerandra frutescens*, a powerful insect repellent for humans. And a seven-year study of Minnesota prairie grasslands, released least year, found that biological diversity helps keep ecosystems healthy.

In the study, scientists found that the grassland plots with the greatest variety of plant species lost far less of their vegetative cover than plots with fewer species of plants during the worst U.S. drought in five decades, and they recovered more quickly. Diversity is "nature's insurance policy against catastrophes," said one of the study's authors, ecologist David Tilman of the University of Minnesota.

On another front, the news in 1994 regarding the nation's freshwater and saltwater fisheries was not encouraging. According to an Environmental Defense Fund report, America's freshwater fish and shellfish are the nation's most endangered forms of wildlife. In reaching that sobering conclusion, the report cited a number of regional declines, including the disappearance of two-thirds of all fish species in the Illinois River and 96 percent of the hickory shad in the Chesapeake Bay.

The Fish and Wildlife Service found nearly 40 percent of the animals and plants protected by the Endangered Species Act are stable or improving.

The situation is no better in U.S. coastal waters. Much of Cape Cod, America's oldest commercial fishery, is now declared by authorities to be off-limits to fishermen because of a scarcity of cod, haddock, and flounder. The jobs of 20,000 U.S. fishermen

in New England are imperiled by the decline of their groundfish catch from 780,000 metric tons in 1965 to 25,000 metric tons in 1993. Last year, the U.S. Commerce Department imposed strict catch limits, then authorized $30 million in emergency aid to the depressed New England industry.

While the declines in coastal fisheries can be attributed to overfishing for the most part, pollution, habitat loss, and dam construction have been the primary causes of the severe depletion of Pacific salmon in the Northwest. Last April, federal authorities banned salmon fishing off Washington State for the first time ever in the nation's history and imposed tough catch limits off Oregon and California in an effort to ensure that enough of the fish survive to perpetuate their species. Four months later, federal officials reclassified Snake River chinook salmon from threatened to endangered, a move caused by the continuing decline of what was once one of the world's greatest chinook runs.

Air

While states stall in complying with federal law, one in four Americans continues to breath unhealthful air.
How safe is the air Americans breathe these days? According to the Environmental Protection Agency (EPA), the number of U.S. citizens living in areas that do not meet the federal ozone standard declined by 37 percent to 54 million in 1992 (the most recent figures available)—the lowest number in two decades of monitoring.

Federal officials also reported that emissions of hydrocarbons—major components of smog—by new automobiles have been reduced by 90 percent since passage of the Clean Air Act in 1970. And EPA data on the release of airborne toxic chemicals by U.S. industry show a 9.4 percent improvement over 1991 and an overall reduction of 35 percent since 1988.

Such reductions represent considerable improvement by any measure. But, experts concede, getting rid of air pollution—particularly the ground-level ozone pollution that cripples urban areas during the warmer months of the year—is a daunting

task. "One out of four Americans still lives in areas where the air is not safe to breathe," observed Mary Marra, director of the National Wildlife Federation's Environmental Quality Division, last year.

The Los Angeles metropolitan area, classified by federal authorities as "extremely polluted," suffers the worst smog levels in the nation. Seven other cities and their suburbs—New York, Chicago, Houston, Milwaukee, Baltimore, Philadelphia, and San Diego—are classified as "severely polluted."

Meanwhile, according to a report released last April by the American Lung Association, some 23 million Americans currently face a "deadly public health threat" because they live in areas where they breathe high concentrations of particulates—tiny airborne particles, especially those produced by fossil fuel combustion, that can impair lung function. The report came on the heels of a 16-year study by Harvard University's School of Public Health, which found that even people living in those regions meeting the existing federal air-quality standards for particulates showed serious health problems related to breathing the tiny particles, including elevated death rates associated with air pollution.

Five years ago, Congress and the Bush administration decided that the bad news about air pollution—100 cities had never complied with the Clean Air Act—so outweighed the good that drastic action was necessary. In a series of amendments to the 1970 act, lawmakers set stiff new standards for reducing urban smog and other air pollution, and they established tough penalties for failure to meet those standards. Yet as last summer came to a close, 10 states had not filed plans that had been due in 1992, and some two dozen states that were required to file plans in 1993 had not yet responded.

A primary focus of the 1990 amendments were automobile emissions, still implicated in half of the nation's air pollution. The law required 22 states to initiate more stringent auto inspection and maintenance programs by January 1, 1995. According to EPA officials, better emissions testing could reduce smog levels by as much as 25 percent. But as the deadline neared, most states were not even close to compliance.

While the sanctions mandated by the 1990 amendments had yet to be enforced, California authorities received an advance look at the potential consequences of continued noncompliance when environmentalists sued the state for failing to meet deadlines in the 1977 version of the Clean Air Act. A federal court ordered EPA to do what the law required: impose a plan to bring the state into compliance. And last February, the agency issued a 2,700-page summary of 100 measures required to bring healthful air to Los Angeles and its surrounding counties by the year 2010. The measures include several innovative pollution fees and clean-air standards.

In another measure mandated by the 1990 amendments, EPA last year issued new rules requiring U.S. chemical plants to cut by 88 percent the amount of toxic substances they emit into the air over the next three years. "This is equivalent to taking about one-fourth of all cars in America off the road," said EPA Administrator Carol Browner.

Scientists reported last summer that the ozone layer high over North America had rebounded from its extremely low level of two winters ago. Not to be confused with ground-level ozone pollution that chokes cities, atmospheric ozone blocks ultraviolet radiation from the sun. Such radiation can lead to skill cancer and other problems.

Experts cautioned, however, that the recovery did not mean people could now safely spend time in the sun without protection. "There's still the long-term decline [in the ozone layer] that's been going on for the past dozen or so years," said Samuel J. Oltmans of The National Oceanic and Atmospheric Administration. "It's not a recovery to levels we saw in the mid-1970s."

Energy

As federal agencies strive to reduce energy consumption, low oil prices negate some conservation tactics. Early last year, environmentalists and U.S. oil company executives found themselves sharing the same concern: low oil prices. As the year began, the average cost of crude oil continued to fall despite high demand during a cold winter, limited reserve capacity among members of OPEC, and the continuing absence from the market of any oil from Iraq. In January, crude oil was selling at a five-year low of $12 a barrel. And OPEC, which once virtually dictated world oil prices by manipulating supplies, failed in two attempts to persuade its members to cut production by just 2 percent.

Environmentalists were dismayed because cheap oil meant a continuing lack of economic incentives to develop or switch to alternative energy sources. Average regular gasoline prices at the pump fell in January to $1.06 per gallon, obliterating the effects of the small energy tax imposed by the federal government the previous October to encourage conservation.

U.S. oil companies were unhappy with the low prices because 15 percent of the 6.6 million barrels per day they are capable of producing comes from wells that are very expensive to operate. Oil executives estimated that they needed prices of about $18 per barrel to keep such wells profitable. By midyear, oil prices had inched upward but still hovered below $17 per barrel—low enough, observed Ed Rothschild of the nonprofit group Citizen Action, to keep people "hooked on oil."

Despite the bargain-basement prices, many electric utilities continued to reduce their reliance on oil in 1994 in order to comply better with federal clean-air laws. Such utilities, along with certain other industries, are capable of switching fuels. But last year, the companies tended to stick with natural gas, which burns cleaner, even when oil would have been cheaper.

In an effort to avoid building expensive new power plants, many electric utilities also turned last year to a source they had long criticized as expensive and unreliable: solar energy. Early in 1994, a consortium of 68 electric utilities—from New York City's Consolidated Edison to San Francisco's Pacific Gas and Electric—were finalizing plans to buy $500 million worth of solar panels during the next six years. The utilities serve 40 percent of the nation's electric customers. Their business, said Scott Sklar, director of the Solar Electric Industries Association, "will allow the solar industry to double its manufacturing capacity."

With the cost of converting sunlight to electricity dropping, electric utilities are increasingly plugging into solar power to avoid building new plants.

As a result, the cost of solar power will continue to decline. In the 1960s, photovoltaic cells cost about $500 per watt. In 1993, a new type of solar panel came on the market at less than $4 per watt. Another was expected to sell for about $2.50 per watt in early 1995. At $2.00, the residential cost of solar power will be about the same price as the current California average of 12 cents per kilowatt hour.

Another boost for alternative energy sources came last year when the Clinton administration ordered federal agencies to reduce their energy consumption by 30 percent (of 1985 levels) in 10 years. A 14-page executive order released in March mandated the use of high-efficiency fluorescent lights, better waste-management practices, and renewable energy sources. The measures could conceivably cut an estimated $1 billion per year from the federal government's $11-billion annual energy bill.

Energy Department Secretary Hazel O'Leary said her agency would take the lead in attaining the savings. She noted, for example, that high-efficiency lighting in the department's Washington, D.C., headquarters reduced the light bill by 60 percent in one year. Switching from ordinary to energy-saving computers, which partially shut down when not in use, could save the government another $125 million a year. "Utilities know it is in their long-term interest to help save energy," said O'Leary.

Another government initiative concerning a source of energy did not produce such seemingly harmonious cooperation. The 1990 Clean Air Act amendments permitted EPA to mandate the addition of oxygenates—additives that raise oxygen content and lower emissions that cause smog—to gasoline in the county's nine most polluted cities. Two such additives, ethanol and methanol, are widely available, and Congress told EPA to decide which one was best.

The competition pitted the nation's oil industry, which makes methanol from natural gas, against the agribusiness establishment, which produces ethanol from grain. The nonprofit American Council for an Energy Efficient Economy, which took no stand on the issue, pointed out that growing corn for ethanol requires fertilizing and high-energy practices that can generate pollution. Nevertheless, EPA ruled in July that at least 30 percent of the additives used must come from ethanol. However, in September, a federal appeals court temporarily blocked the EPA decision while it considers oil-industry legal challenges.

Water

Despite grim news about pollution problems, U.S. lawmakers postpone strengthening the Clean Water Act. As Congress began considering reauthorization of the Clean Water Act early last year, EPA officials were reporting the results of their most recent surveys of the health of America's waterways. According to EPA, "Thirty percent of [the nation's] rivers, 42 percent of lakes and 32 percent of estuaries continue to be degraded, mainly by silt and nutrients from farm and urban runoff, combined sewer overflows and municipal sewage."

In summary, EPA Administrator Carol Browner said last spring that 22 years after the law was originally passed, an estimated 40 percent of the country's freshwater is still unusable in terms of public health and safety. While Americans have spent $260 billion in the past two decades on new sewage-treatment plants, water treatment remains substandard in 1,100 U.S. cities.

EPA also reported a 12 percent increase in the amount of toxic chemicals discharged into U.S. surface waters. And a General Accounting Office (GAO) study last April raised questions about EPA's ability to enforce existing controls on toxic-substance releases in water. In an examination of 236 companies with permits to discharge toxic chemicals, GAO found that 77 percent of the chemicals discharged were not listed on the permits. In many cases, said GAO, the unreported chemicals were recognized as health risks.

Despite such grim news, opposition was strong in 1994 against any attempts to add more regulatory teeth to the Clean Water Act during the reauthorization process. Special interests, represented by groups like the American Farm Bureau Federation, lobbied lawmakers against passing more stringent controls of farm runoff pollution, which along with stormwater runoff from city streets accounts for half of all pollutants in U.S. surface waters, according to EPA.

Environmentalists, on the other hand, continued to push lawmakers to impose stricter controls on runoff and chemical contamination. "We cannot in good conscience sit back and allow special-interest groups to succeed in weakening one of the nation's most important public health laws," said Sharon Newsome, NWF vice president for resources conservation. After considerable debate on the subject, Congress took no action in 1994 on the Clean Water Act.

Federal lawmakers also failed last year to reauthorize two other important measures that affect the quality of the nation's water: the Safe Drinking Water Act and the Superfund law. The latter was created in 1980 to clean up

the nation's worst abandoned toxic-waste sites, many of which are leaking dangerous chemicals into water supplies. However, because of bureaucratic tangles, only about 200 of the 1,200 priority sites targeted since 1980 have been cleaned up. With funding scheduled to run out in 1995, Congress proposed a number of measures to reform the law.

Under the proposals, EPA would set more flexible standards for cleanup of Superfund sites, based on future intended use of the land. Environmentalists welcomed such changes while arguing that Congress should maintain the law's controversial liability system. "This ensures polluters, not taxpayers, will pay for cleaning up a mess they created," said NWF legislative representative Patricia Williams.

Regarding the Safe Drinking Water Act, which technically expired in 1991, lawmakers and the Clinton administration agreed last year to exempt some small municipalities from meeting all of its mandates. They also agreed to ease rules on some larger communities.

Reluctantly supporting the agreements, environmentalists pointed to recent incidents of contamination of municipal water supplies. "There are 200,000 public water systems in this country, a substantial percentage of which are basket cases that cannot even meet the most basic microbiological standards," said Eric Olson, an attorney with the Natural Resources Defense Council.

Meanwhile, last May, the U.S. Supreme Court resolved a dispute over whether the states have authority under the Clean Water Act to protect not only the quality but also the *quantity* of water that flows in streams and rivers. The case involved an electric utility in Washington State that requested a permit to build a hydroelectric dam. The state, however, required that as a condition of building the dam, the utility must maintain a minimum water flow to protect Pacific salmon and steelhead trout. The utility filed suit, charging that the law mentions only water quality, not quantity.

Writing for the seven to two majority of the Supreme Court, Justice Sandra Day O'Connor said that the attempt by the utility to separate the issue of water quality and quantity was an "artificial distinction." The law was intended, she wrote, "to protect the physical and biological integrity of water," and she said it was clear that "a sufficient lowering of quantity in a body of water could destroy all of its designated uses."

Soil

Can farmers continue to reduce soil erosion and improve wildlife habitat in many regions of the country? Last summer, the U.S. Soil Conser-vation Service reported that American farmers were well on the way to reducing soil losses on at least 140 million acres of the country's most erodible farmland. A decade ago, farmers were losing an estimated average of 7.4 tons per acre to erosion. The latest data show that figure has been reduced to 5.6 tons per acre in 1994. A part of the success was due to the Conservation Reserve Program (CRP), established in the 1985 Farm Bill that has paid thousands of farmers to remove environmentally sensitive land from production and plant it with grass or trees to prevent erosion.

Where the CRP has worked effectively, wildlife has benefited. According to research by federal scientists, 16 grassland bird species that have been in long-term decline are more abundant on CRP land than in surrounding areas. And officials in North Dakota and South Dakota estimate that their states are producing 1 million more ducks each year because of CRP.

Ironically, farmers in nine Midwest states along the Missouri and Mississippi Rivers struggled during 1994 not with erosion but with sedimentation—millions of tons of sand and other debris left on their cropland by the flood waters of 1993. The situation was made worse by a levee system that contained and accelerated the flood waters, which then picked up more sediment before finally breaking through the levee constraints. Last

May, a committee of federal experts recommended levees along the Missouri River be moved 2,000 feet back from the water—a recommendation that if implemented could result in reduced damage from future flooding to farms and wetlands.

In late 1993, the National Wildlife Federation and other groups persuaded lawmakers to make funds available for purchase of floodplain-damaged lands from farmers for restoration back to wetlands. Since then, nearly 100,000 acres have been offered to the program.

The federal flood-relief effort also included funds to help move homes and businesses away from floodplains. By last summer, more than 40 towns had asked for relocation assistance and two towns elected to move entirely. Instead of managing the river, said Mayor Rodney Horrigan of Chelsea, Iowa, "We're trying to manage the people and let the river go where it wants to."

On another front, Interior Department Secretary Bruce Babbitt launched a third attempt since taking office to reform the management of 270 million acres of federal rangeland in the West. Bargain grazing fees of $1.86 per month per animal unit (a cow and calf or equivalent) have for decades encouraged overgrazing on public lands.

A National Wildlife Federation study of public lands in the West found that livestock grazing helped push scores of species to the brink of extinction.

A National Wildlife Federation report released last June, concluded that the degradation of habitat by overgrazing has driven scores of wildlife species closer to extinction. The one-year study found that practices permitted by the U.S. Forest Service and the Bureau of Land Management have done significant

harm to the habitat of 76 fish and wildlife species that are either protected or candidates for protection under the Endangered Species Act. "These species are environmental barometers, providing a warning that native ecosystems across the West are unraveling," the study concluded.

After two earlier attempts at rangeland reform were beaten back by western legislators, Babbitt proposed a top fee of $3.96 per unit instead of the $4.28 he proposed earlier. He also proposed creating regional advisory councils to regulate grazing lands in different regions according to broad federal guidelines. Babbitt said his proposal would bring "significant reforms to the management of our public lands with substantially greater input from westerners." But NWF President Jay D. Hair called the proposal "a watered-down version of reform. The old politics of the West kept major reform measures off the table."

Under mandates of the 1872 federal Mining Law, Secretary Babbitt also was forced last year to transfer ownership of 2,000 acres of federal land in Nevada to a Canadian mining company for a purchase price of $10,000, even though the site holds an estimated $10 billion worth of gold deposits. It represented one of more than 600 applications for land still pending by mine operators who each year take vast amounts of hardrock minerals from public lands without paying royalties to taxpayers.

Last year, Congress remained deadlocked over a new law to replace the 1872 act. While the House passed a bill that would charge an 8 percent royalty to miners and impose new environmental clean-up measures, the Senate's bill imposed only a 2 percent royalty and no restoration measures. "It's an indefensible system that Congress would never enact today—no one would even propose it—yet Congress can't muster the votes to uproot it," read a *Washington Post* editorial last September.

Forests

While the old-growth dispute simmers in the Northwest, a plan to manage Northeast forests is unveiled. The Clinton administration took a long-overdue step in 1994 in addressing the deadlock between the timber industry and conservationists over the future of remaining old-growth forests in the Pacific Northwest. A plan that the administration proposed in 1993 to end the conflict would permit logging of much of the old growth while preserving almost all of the habitat of the region's imperiled wildlife species. Last year, federal officials began to implement portions of that plan.

The proposed plan represented one of the administration's first attempts to replace species-specific recovery plans with ecosystem management as the basis for protection of natural resources. The plan took into account the fact that 40 of the 1,500 species that share old-growth habitat are considered threatened or endangered. It also represented the first test of President Clinton's determination to find consensus among opponents in environmental issues.

Not until last June, however, did a federal judge lift a three-year-old injunction that prohibited all logging on Pacific Northwest federal lands, pending development of a reasonable plan to protect the threatened northern spotted owl, as required by federal law. The judge's action allowed the U.S. Forest Service to begin preparing timber sales that met the criteria of the plan. But by that time, the timber industry and environmentalists had filed lawsuits to challenge the plan.

The environmentalists insisted that the spotted owl, as a conference of forest biologists reaffirmed last February, "is approaching the extinction threshold" and may not be able to survive "any additional habitat loss." That conclusion was supported last summer when federal officials rejected a timber-industry petition to remove the spotted owl from the state list of threatened species in California. Environmentalists were also concerned that the Clinton plan may not

adequately protect certain runs of endangered and threatened Pacific salmon and other Northwest species.

On the other side of the country, members of the Northern Forest Lands Council, an independent body created by Congress in 1990, sought last year to avoid the kinds of timber-related conflicts that have plagued the Northwest. The council's mandate is to develop a consensus on management of another major woodland area: the 26-million-acre Northern Forest, which flourishes from Maine to New York State.

After taking into account the goal of sustaining the woodlands along with the needs of private landowners who hold 85 percent of the forest, the hundreds of wildlife species that inhabit it, the people whose livelihoods depend upon timbering, and the 70 million others who live within an eight-hour drive of the woodlands, the council recommended 37 incentives and regulations. Predictably, every measure generated controversy among competing interests.

An NWF study in 1994 of timbering in the region found that an increasing incidence of untracked exports of raw logs threatens both the ecological and economic health of the Northern Forest. "A significant portion of the highest-quality logs, which might have been used to create and sustain local jobs through processing and manufacturing, are being shipped out of the region," said NWF resource economist Eric Palola.

To prevent further loss of wetlands in the Southeast, the Environmental Protection Agency banned companies from draining marshes to plant timber.

In another of the nation's major forested areas, the Southeast, EPA moved in 1994 to close a loophole in environmental law that has contributed to the destruction of millions of acres of wetlands. The Clean Water

Act regulates activities in the country's remaining 100 million acres of wetlands, one-third of which are located in the Southeast. In the past, the law has permitted so-called "normal" forestry practices in wetlands, and for years timber companies have considered clearing and draining wetlands a normal process for creating pine plantations. Last February, however, the EPA specifically banned such practices.

Meanwhile, last year the U.S. Forest Service appeared to intensify its efforts to transform itself, as promised by the Clinton administration, from a timber-selling agency to one that manages ecosystems. For decades, the operations of individual national forests have been funded largely according to how much timber they sold. To facilitate sales, the agency has built at taxpayers' expense more than 370,000 miles of logging roads in public forests—a network that is eight times the size of the federal Interstate Highway System—frequently at costs that far exceeded the value of the timber sold.

"The Forest Service was once the conservation leader in the United States and the world," said its new chief, Jack Ward Thomas, last year. "It needs to get back to its roots." One example of the change in priorities: In 1994, the Olympic National Forest did not sell timber with the same emphasis it had in the past, but had a budget of nearly $3 million to spend on watershed studies and restoration.

Quality of Life

Research uncovers potential new threats to human health from an ominous class of toxic pollutants. If many Americans expressed distress in 1994 about the effects of environmental contaminants on the quality of their lives, they had good reason for concern. Last year, they were confronted with several new studies about how an ominous class of pollutants may be affecting human health.

The pollutants in question often are found in traces so small that they are hard to measure and act so slowly that their full impact may not be known for generations. Instead of directly poisoning, these compounds may imitate the natural hormones of living creatures well enough to take over and then sabotage the many functions the hormones control. "When pollutants that mimic hormones (such as estrogen) trick the body into accepting them as natural hormones, the health effects can range from subtle cellular changes to permanent disfiguring impairments," said a National Wildlife Federation report last spring.

The harm such compounds may do to human health is difficult for researchers to prove. What's more, because each substance helps produce corporate profits, every shred of evidence of a chemical's dangers often is countered by massive campaigns of denial by industry.

Consider chlorine, which is used to whiten clothing and paper, purify water, and help refrigerate products—and which plays a role in the manufacture of many pesticides and plastics. When chlorine combines with carbon, the two can form some of the most persistent toxic substances on Earth, including PCBs, DDT, and dioxin. A draft EPA study released in 1994 linked dioxin not only to cancer but also to disastrous hormonal and immunological changes in people and wildlife.

The threat to wildlife was documented last year by the U.S.-Canadian International Joint Commission, which reported that of 42 compounds known to be afflicting the reproductive or hormonal systems of animals in the Great Lakes regions, more than half "contains chlorine as an essential ingredient."

The threat to humans has proven more difficult to document, and federal authorities are increasing efforts to study the problem. Last January, the National Institutes of Health in Maryland devoted an entire conference to the topic. That same month, as Congress began considering renewal of the Clean Water Act, the Clinton administration proposed developing "a national strategy for substituting, reducing or prohibiting the use of chlorine and chlorinated compounds." The Chlorine Chemistry Council responded by charging that critics of their products were not using "a sound-science approach to decision making."

Although often confused by such claims and counterclaims, Americans held firm in 1994 to their concern about threats to their health. The Harvard Center for Risk Analysis reported last year that 66 percent of Americans interviewed agreed that "the government is not doing enough to protect people from environmental pollution." Ninety-four percent wanted to know the estimated costs and benefits of new clean-up regulations.

For many of the nation's minorities and economically disadvantaged communities, concerns continued to grow in 1994 about evidence of environmental injustice: documentation that many of the worst pollution sources and hazardous-waste sites are located in the poorest communities with the highest proportions of minorities. In response to a rising call for environmental justice, the Clinton administration agreed last year to investigate the possibility that the sit-

ing of hazardous waste plants in minority communities in Louisiana and Mississippi violated the residents' civil rights.

"It's clear," said EPA Administrator Carol Browner, "that low-income and minority communities have been asked to bear a disproportionate burden of the country's industrial lifestyle. We have to incorporate environmental justice into everything we do." An executive order from President Clinton in February to federal agencies made that concern national policy.

The public's desire for an accounting of costs and benefits, as demonstrated in the Harvard survey, may reflect not only their confusion over environmental hazards, but also of the claims of many industries that correcting pollution problems will cost too much and eliminate jobs. In 1994, the Commerce Department began implementing a form of accounting that may not be to the liking of such industries.

The program represents a recalculation of the country's total output of goods and services by treating natural resources as assets and subtracting the value of resources consumed from the so-called Gross Domestic Product. This "green accounting" focused on minerals last year, will move on to renewable surface resources in 1995, and to pollution costs in 1996. When done, the project should provide a better picture of the sustainability of U.S. economic activity—a dollar-and-cents measure of the nation's true quality of life.

THOUGHT EVOKER

- Describe the positive environmental advances that have taken place during the past year regarding ecological conditions in the United States.

National Wildlife's *annual Environmental Quality Review is a subjective analysis of the state of the nation's natural resources. The information included in each section is based on personal interviews, news reports, and the most current scientific studies.*

24

Serotonin, Motor Activity, and Depression-Related Disorders

Clues to the origin and treatment of depression and obsessive-compulsive disorders can be found in the role of serotonin neurons in the brain

By Barry L. Jacobs

Prozac, Zoloft, and Paxil are drugs that have been widely celebrated for their effectiveness in the treatment of depression and obsessive-compulsive disorders. The popular press has also made much of Prozac's ability to alleviate minor personality disorders such as shyness or lack of popularity. The glamorous success of these drugs has even inspired some writers to propose that we are at the threshold of a new era reminiscent of Aldous Huxley's *Brave New World,* in which one's day-to-day emotions can be fine-tuned by simply taking a pill. Yet for all the public attention that has been focused on the apparent benefits of Prozac-like drugs, the fundamental players in this story—the cells and the chemicals in the brain modified by these drugs—have been largely ignored.

This is partly a consequence of the complexity of the nervous system and the fact that so little is known about *how* the activity of cells in the brain translates into mood or behavior. We do know that Prozac-like drugs work by altering the function of neurons that release the signaling chemical (neurotransmitter) serotonin. Serotonin has been implicated in a broad range of behavioral disorders involving the sleep cycle, eating, the sex drive, and mood. Prozac-like drugs prevent a neuron from taking serotonin back into the cell. Hence Prozac and related drugs are collectively

Barry L. Jacobs is professor and director of the program in neuroscience at Princeton University. Address: Department of Psychology, Green Hall, Princeton University, Princeton, NJ 08544.

known as selective serotonin reuptake inhibitors, or SSRIs. In principle, blocking the reuptake of serotonin should result in a higher level of activity in any part of the nervous system that uses serotonin as a chemical signal between cells. The long-term effects of these drugs on the function of a serotonin-based network of neurons, however, are simply not known.

My colleagues and I have attempted to understand the role of serotonin in animal physiology and behavior by looking at the activity of the serotonin neurons themselves. For more than 10 years at Princeton University, Casimir Fornal and I have been studying the factors that control the activity of serotonin neurons in the brain. I believe these studies provide the linchpin for understanding depression and obsessive-compulsive disorders and their treatment with therapeutic drugs. Our work provides some unique and unexpected perspectives on these illnesses and will serve, we hope, to open new avenues of clinical research.

Serotonin, Drugs, and Depression

Communication between neurons is mediated by the release of small packets of chemicals into the tiny gap, the synapse, that separates one neuron from another. The brain uses a surprisingly large number of these chemical neurotransmitters, perhaps as many as 100. However, the preponderance of the work is done by four chemicals that act in a simple and rapid manner: glutamate and aspartate (both of which excite neurons) and gamma-aminobutyric acid

(GABA) and glycine (both of which inhibit neurons). Other neurotransmitters, such as serotonin, norepinephrine, and dopamine are somewhat different. They can produce excitation *or* inhibition, often act over a longer time scale, and tend to work in concert with one of the four chemical workhorses in the brain. Hence they are also referred to as neuromodulators.

Even though serotonin, norepinephrine, and dopamine may be considered to be comparatively minor players in the overall function of the brain, they appear to be major culprits in some of the most common brain disorders: schizophrenia, depression, and Parkinson's disease. It is interesting to observe that glutamate, aspartate, GABA, and glycine are generally not centrally involved in psychiatric or neurological illnesses. It may be the case that a primary dysfunction of these systems is incompatible with sustaining life.

Serotonin's chemical name is 5-hydroxytryptamine, which derives from the fact that it is synthesized from the amino acid L-tryptophan. After a meal, foods are broken down into their constituent amino acids, including tryptophan, and then transported throughout the body by the circulatory system. Once tryptophan is carried into the brain and into certain neurons, it is converted into serotonin by two enzymatic steps.

Serotonin's actions in the synapse are terminated primarily by its being taken back into the neuron that released it. From that point, it is either recycled for reuse as a neurotransmitter or broken down into its metabolic

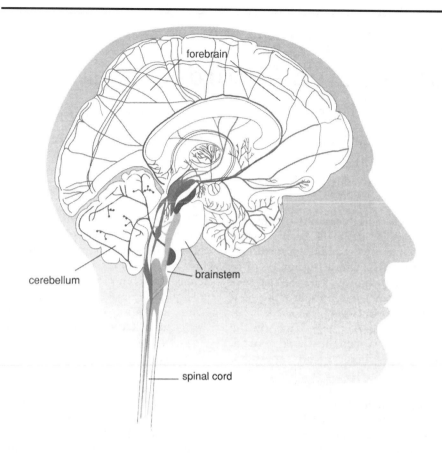

forebrain

cerebellum

brainstem

spinal cord

Clusters of serotonin neurons located in the raphe nuclei within the brainstem project throughout most regions of the forebrain, brainstem, cerebellum, and spinal cord. The widespread distribution of the serotonin-releasing fibers at least partially accounts for the system's influence over such basic functions as the sleep cycle, the sex drive, eating, body temperature, cardiovascular activity, respiration, mood, and aggression. (Adapted from the work of Efrain Azmitia, New York University.)

by-products and transported out of the brain. With this basic understanding of serotonin neurotransmission, we can begin to understand the mechanisms of action of antidepressant drugs.

One of the earliest antidepressant drugs, iproniazid, elevates the level of a number of brain chemicals by inhibiting the action of an enzyme, monoamine oxidase, involved in their catabolism. For example, monoamine oxidase inhibitors (MAOIs) block the catabolism of serotonin into its metabolite, 5-hydroxyindole acetic acid (5-HIAA), leading to a buildup of serotonin in the brain. Unfortunately, because monoamine oxidase catabolizes a number of brain chemicals (including norepinephrine and dopamine), there are a number of side-effects associated with these drugs. Some interactive toxicity of MAOIs is also a major drawback for their use in the treatment of depression.

Tricyclic antidepressants (so named because of their three-ringed chemical structure) do not share the interactive toxicity of MAOIs. Tricyclics such as imipramine act to block the reuptake of serotonin from the synapse back into the neuron that released it. In a sense, this floods the synapse with serotonin. These drugs are quite effective in treating depression, but they also induce some unpleasant side-effects, such as constipation, headache, and dry mouth. This may be due to the fact that tricyclic antidepressants not only block the reuptake of serotonin, but also exert similar effects on norepinephrine and dopamine.

The obvious benefit of the selective serotonin reuptake inhibitors is that their action is effectively limited to the reuptake of serotonin. This probably accounts for the fewer side-effects experienced by people taking SSRIs. Like other antidepressant drugs, the

SSRIs have a therapeutic lag. They typically require 4 to 6 weeks to exert their full effects. Claude DeMontigny and his colleagues at McGill University have suggested that one of the consequences of increasing the levels of serotonin in the brain is a compensatory feedback inhibition that decreases the discharge of brain serotonergic neurons. This results in a "zero sum game" in which there is no net increase in functional serotonin. However, with continuous exposure to serotonin, the receptors mediating this feedback inhibition (the 5-HT_{1A} receptor) become desensitized. It is hypothesized that after several weeks this results in progressively less feedback, increased serotonergic neurotransmission, and clinical improvement.

Behavior of Serotonin Neurons

Essentially all of the serotonin-based activity in the brain arises from neurons that are located within cell clusters known as the raphe nuclei. These clusters of serotonin neurons are located in the brainstem, the most primitive part of the brain. It is not surprising, then, that serotonin appears to be involved in some fundamental aspects of physiology and behavior, ranging from the control of body temperature, cardiovascular activity, and respiration to involvement in such behaviors as aggression, eating, and sleeping.

The broad range of physiology and behavior associated with serotonin's actions is at least partly attributable to the widespread distribution of serotonin-containing nerve-fiber terminals that arise from the raphe nuclei. Indeed, the branching of the serotonin network comprises the most expansive neurochemical system in the brain. Serotonin neurons project fibers to virtually all parts of the central nervous system, from the various layers of the cerebral cortex down to the tip of the spinal cord.

Since the raphe nuclei contain only a few hundred thousand neurons, and the brains of large mammals (cats, monkeys, and people, for example) contain hundreds of billions of neurons, the serotonin neurons constitute less than one-millionth of the total population of neurons in the brain. Despite being so vastly outnumbered, serotonin neurons have immense importance: *Each one* exerts an influence over as many as 500,000 target neurons. In this light the activity of individual serotonin neurons bears a closer look.

The behavior of any neuron is typically measured by its electrical activity. One of the more remarkable electrical phenomena is a cell-wide discharge called an action potential, or spike. Action potentials result from the movement of charged particles (potassium and sodium ions) into and out of the cell through specialized channels in the membrane. Action potentials are an important aspect of a neuron's behavior because the rate and pattern of their occurrence is thought to encode information that is conveyed to other cells. They are central to our story because, in the case of serotonin neurons, each electrical discharge results in the release of a small packet of serotonin from the cell, which in turn alters the activity of target cells bearing serotonin receptors. (At this writing, at least 14 types of serotonin receptors are known, each of which contributes to the diverse effects of serotonin throughout the brain.)

Serotonin neurons have a characteristic discharge pattern that distinguishes them from most other cells in the brain. They are relatively regular, exhibiting a slow and steady generation of spikes. Serotonin neurons retain this rhythmic pattern even if they are removed from the brain and isolated in a dish, suggesting that their clocklike regularity is intrinsic to the individual neurons.

One of the first significant discoveries about the behavior of serotonin neurons in the brain was that the rate of these discharges was dramatically altered during different levels of behavioral arousal. When an animal is quiet but awake, the typical serotonin neuron discharges at about 3 spikes per second. As the animal becomes drowsy and enters a phase known as slow-wave sleep, the number of spikes gradually declines. During rapid-eye movement (REM) sleep, which is associated with dreaming in

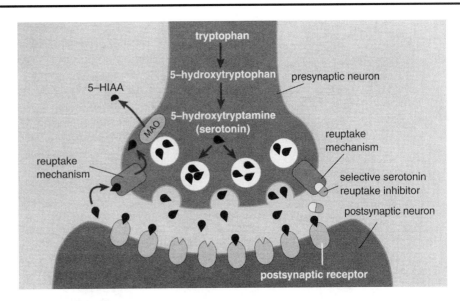

Activity at the site of a serotonin synapse between two neurons involves both the release and reuptake of the neurotransmitter. Serotonin is synthesized from the amino acid tryptophan and sequestered within small packets or vesicles. The arrival of an action potential at the presynaptic terminal results in the release of serotonin from the vesicles into the synaptic gap. The binding of serotonin to specialized receptors on the postsynaptic neuron produces a reaction that alters the electrical and chemical activity of the receiving neuron. Serotonin is removed from the synaptic cleft by another mechanism that takes the neurotransmitter back into the presynaptic terminal, where it is reused or degraded into its primary metabolite, 5-hydroxyindole acetic acid (5-HIAA). Antidepressant drugs such as iproniazid inhibit the enzyme monoamine oxidase (MAO), which normally acts to degrade serotonin. A new generation of antidepressant drugs, including selective serotonin reuptake inhibitors such as Prozac, allow serotonin to remain in the synaptic cleft by blocking its reuptake into the presynaptic neuron. Consequently, the functional activity of serotonin is increased.

human beings, the serotonin neurons fall completely silent. In anticipation of waking, however, the neuronal activity returns to its basal level of 3 spikes per second. When an animal is aroused or in an active waking state, the discharge rate may increase to 4 or 5 spikes per second.

Since early experimental studies had linked serotonin to so many behavioral and physiological processes, one of our first priorities was to examine the activity of serotonin neurons in animals exposed to a wide variety of conditions. We chose to perform these studies on cats since their brains and their raphe nuclei have been described in considerable detail. While we recorded the electrical activity of individual serotonin neurons in the brain, we exposed cats to various stressors such as loud noise, physical restraint, a natural enemy (a dog), a heated environment, a fever-inducing agent, drug-induced changes in blood pressure, and insulin-induced changes in plasma glucose levels. These condi-

tions would be stressful to any animal and, not surprisingly, each of these stressors resulted in dramatic changes in the animal's behavior and activated its emergency defenses, including certain parts of its autonomic nervous system. Remarkably, however, none of these conditions significantly changed the activity of the serotonin neurons beyond that seen during a spontaneous active state. The results were perplexing. If these powerful stimuli could not perturb a serotonin neuron, what would?

The variable activity of serotonin neurons during different stages of the sleep-wake-arousal cycle provided a clue. Recall that serotonin neurons are silent during REM sleep. One of the fundamental features of REM sleep is paralysis of the major muscles of the body. This is achieved by inhibiting the neurons that control the tone of the body's antigravity muscles. Might there be a relationship between the paralysis that takes place during REM sleep and the silence of the serotonin neurons?

This question was addressed with a relatively simple experiment. The destruction of a discrete part of a cat's brainstem produces an animal that by all criteria appears to enter REM sleep. However, antigravity muscle tone is present in these animals, and consequently they are capable of movement and even coordinated locomotion. (This condition has also been observed in some people who have experienced traumatic brain injuries.)

When such a cat is awake or in slow-wave sleep, the activity of its serotonin neurons is similar to that of a normal cat. When these animals enter REM sleep, however, instead of falling silent, the activity of the serotonin neurons increases. Cats that display the greatest amount of muscle tone and overt behavior during REM sleep also have the most active serotonin neurons. In some instances, the activity of their serotonin neurons

active waking   quiet waking   slow-wave sleep   REM sleep

serotonin neuronal activity

5 seconds

Activity of a serotonin neuron varies across an animal's sleep-wake-arousal cycle. The neuron's clocklike generation of action potentials ranges from 5 per second in the active waking state to about 0 per second during rapid-eye-movement (REM) sleep. The absence of muscle tone during normal REM sleep (when serotonin neurons are inactive) suggests a possible link between the activity of serotonin neurons and the facilitation of muscle activity.

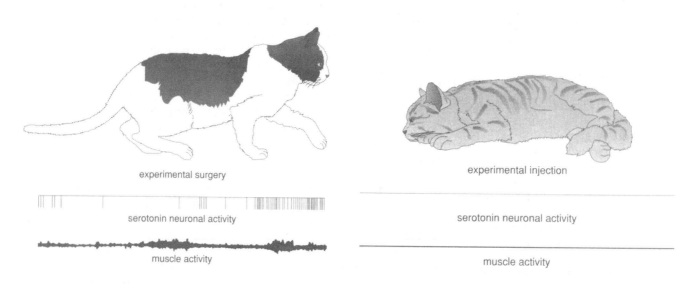

experimental surgery

serotonin neuronal activity

muscle activity

experimental injection

serotonin neuronal activity

muscle activity

Experimental destruction of a discrete region in the brainstem produces an animal that reestablishes muscle tone and movement when it enters REM sleep. Unlike normal REM sleep, the animal's serotonin neurons are active and it is capable of moving about its environment. The experiment supports the hypothesis that serotonin neurons facilitate muscle activity.

Experimental injection of a drug that suppresses the activity of serotonin neurons in the brainstem produces a condition reciprocal to REM sleep. The cat is awake but paralyzed as it would be in REM sleep. The experiment further supports the hypothesis that serotonin neurons facilitate muscle activity.

reaches the same level as that seen during the normal waking state.

Another experiment provided further evidence for the role of serotonin neurons. When a drug that mimics the action of the neurotransmitter acetylcholine is injected into the same region of the brainstem as in the previous experiment, a condition somewhat reciprocal to normal REM sleep can be produced. These animals are awake, as demonstrated by their ability to visually track a moving object, but they are otherwise paralyzed. As in the normal animal in REM sleep, the serotonin neurons in these animals are completely silent. In association with our earlier studies, these results suggest that we are closing in on at least one of the roles played by serotonin in the brain: There is clearly a strong relationship between the activity of serotonin neurons and the body's motor activity.

Interestingly, some serotonin neurons tend to become active just *before* a movement begins. Their activity may also occasionally synchronize with a specific phase of the movement—discharging most, for example, during a particular aspect of the quadrupedal stepping cycle. Moreover, the rate of the spike discharge often increases linearly with increases in the rate or strength of a movement, such as an increase in running speed or the depth of respiration.

One final observation provides a noteworthy clue to the function of serotonin neurons. When an animal is presented with a strong or novel stimulus, such as a sudden loud noise, it often suppresses all ongoing behavior, such as walking or grooming, and turns toward the stimulus. This orienting is essentially a "what is it?" response. In such instances serotonin neurons fall completely silent for several seconds and then resume their normal activity.

Anatomical evidence supports these observations about the activity of serotonin neurons. For one thing, serotonin neurons preferentially make contacts with neurons that are involved in tonic and gross motor

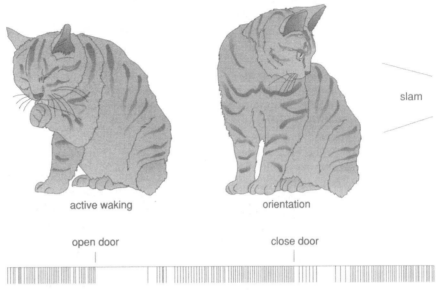

serotonin neuronal activity

Serotonin neurons briefly fall silent when an animal's attention is drawn to a novel stimulus, such as a loud noise made by a door opening or closing. At such times the animal stops all ongoing behavior, such as walking or grooming, and orients to the stimulus in a "what is it?" response. When serotonin neurons are inactive, motor output is disfacilitated and sensory-information processing is disinhibited.

treadmill locomotion

serotonin neuronal activity

Serotonin neurons increase their activity when an animal engages in any of a variety of repetitive behaviors such as chewing food or running on a treadmill. The rate of the action potentials also increases with the rate of the repetitive activity. Here the activity of a serotonin neuron is synchronized with a particular phase of the animal's gait.

functions, such as those that control the torso and limbs. Reciprocally, serotonin neurons tend not to make connections with neurons that carry out episodic behavior and fine movements, such as those neurons that control the eyes or the fingers.

Our observations of the activity of serotonin neurons during different aspects of an animal's behavior lead us to conclude that the primary function of the brain serotonin system is to prime and facilitate gross motor output in both tonic and repetitive modes. At the same time, the system acts to inhibit sensory-information processing while coordinating autonomic and neuroendocrine functions with the specific demands of the motor activity. When the serotonin system is not active (for example, during an orientation response), the rela-

tions are reversed: Motor output is disfacilitated, and sensory-information processing is disinhibited.

Brain Cells and Mental Disorders

Although we are far from understanding the precise neural mechanisms involved in the manifestation of any mental illness, a number of studies have linked serotonin to depression. One of the most notable findings is that the major metabolite of serotonin (5-HIAA) appears to be significantly reduced in the cerebrospinal fluid of suicidally depressed patients. Our own studies suggest that serotonin neurons may be centrally involved in the physiological abnormality that underlies depression-related disorders. Recall that serotonin neurons appear to play crucial roles in facilitat-

ing tonic motor actions and inhibiting sensory-information processing. If an animal's serotonin neurons are responding abnormally, such that the rate or pattern of their activity is modified, then one might expect that both motor functions and sensory-information processing would be impaired.

Depression is frequently associated with motor retardation and cognitive impairment. If serotonin neurons are facilitating tonic motor activity, then it should not be surprising that depressed patients feel listless and often appear to require enormous effort merely to raise themselves out of bed. Inappropriate activity during sensory information processing might also account for the lapses of memory and the general lack of interest in the environment experienced by depressed patients. It might also be worth noting here that the well-known efficacy of REM-sleep deprivation for treating depression is at least partly dependent on serotonin. Since serotonin neurons are usually silent during REM sleep, depriving an animal of REM sleep maintains a generally higher level of activity in the system. Preliminary research in my laboratory suggests that the deprivation of REM sleep also increases the activity of serotonin neurons when the animal is in the awake state.

The activity of serotonin neurons may also be central to the manifestation of obsessive-compulsive disorders. Since our results show that repetitive motor acts increase serotonin neuronal activity, patients with this disorder may be engaging in repetitive rituals such as hand washing or pacing as a means of self-medication. In other words, they have learned to activate their brain serotonin system in order to derive some benefit or rewarding effect, perhaps the reduction of anxiety. Since the compulsive acts tend to be repeated, often to the point of becoming continuous, such activity may provide an almost limitless supply of serotonin to the brain. (The same may also be true for repetitive obsessional thoughts, but this is obviously difficult to test in animals.) Treating obsessive-compulsive disorders with a selective sero-

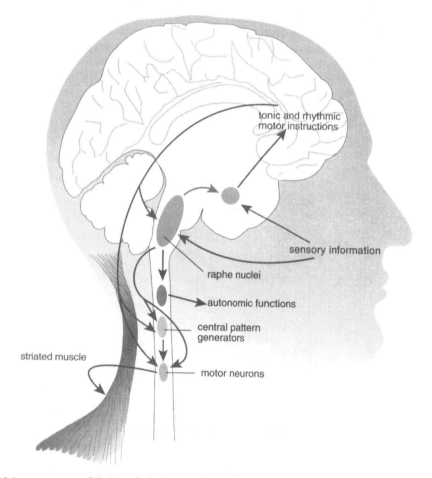

Major components of the hypothesized motor functions of the serotonin system are distributed in different parts of the nervous system. Excitatory input from the forebrain influences the activity of serotonin neurons in the brainstem, as well as central-pattern-generating neurons and motor neurons in the brainstem and spinal cord during tonic or rhythmic motor activity. Serotonin neurons in the brainstem serve several functions: They facilitate central pattern generator and motor-neuron activity, coordinate autonomic and neuroendocrine output with motor output, and inhibit sensory-information processing.

tonin reuptake inhibitor ultimately accomplishes the same neurochemical endpoint, thus allowing these people to disengage from time-consuming, socially unacceptable, and often physically harmful behavior.

Our studies suggest that regular motor activity may be important in the treatment of affective disorders. For example, if there is a deficiency of serotonin in some forms of depression, then an increase in tonic motor activity or some form of repetitive motor task, such as riding a bicycle or jogging, may help to relieve the depression. Indeed, there are various reports that jogging and other forms of exercise have salutary effects for depressed patients. This does not mean that exercise is a panacea for depressive disorders. Since the long-term effects of exercise on brain serotonin levels are not known, the benefits may prove to be transient. On the other hand, exercise may be an important adjunct to drug treatments and may permit a reduction in the required drug dosage.

Finally, there is another potentially productive avenue for drug intervention in the clinic. It is well established that the activity of serotonin neurons is under negative feedback control, in which the released serotonin molecules bind to the so-called autoreceptor on the releasing cell and act to inhibit the cell's activity. Because of this the administration of a metabolic precursor of serotonin, such as the amino acid *L*-tryptophan, cannot significantly elevate the synaptic levels of brain serotonin since neurons compensate by decreasing their activity through the negative-feedback mechanism. However, if *L*-tryptophan treatments are combined with low doses of an autoreceptor-blocking agent, the concentration of serotonin in the synapse might be increased without the use of serotonin reuptake inhibitors. This might prove helpful to patients who have adverse reactions to these drugs and might also circumvent the therapeutic lag of 4 to 6 weeks typically associated with antidepressants. However, because the feedback mechanism is positively correlated with the rate of serotonin neuronal activity, our results suggest that

drugs that block the autoreceptor might be ineffective in quiescent, lethargic, or somnolent patients. Conversely, these drugs may be most effective in active patients or those activated by artificial means.

Conclusion

Our research raises the issue of why the manipulation of a system that is primarily a modulator of motor activity has profound effects on mood. Aside from recognizing that the raphe nuclei are connected to regions of the brain that are known to be involved in the emotions (such as the limbic system), it is worth noting that a common organizational plan underlies the distribution of serotonin cell bodies and fiber terminals in essentially all vertebrate brains. This implies that the system has been conserved through evolution and suggests that there may be some adaptive significance to linking mood and motor activity.

Consider the following possibility. We know that emotions play a role in allowing an animal to withdraw from an ongoing sequence of activities to consider alternative paths. When something bad (perhaps even life threatening) transpires, it seems reasonable to suppress motor activity and to contemplate the available options. To put it another way: If something negative has happened in one's world, it might be counterproductive, or even dangerous, to explore and engage the environment. The most adaptive response is to withdraw and ruminate. In this light, emotions act at a higher level of complexity in the service of effective motor behavior. When one's mood is bright and expansive, on the other hand, it may be profitable to explore new

options. Wide mood swings may allow an exploration of a broader spectrum of perspectives and thus may be related to the well-documented relationship between mood disorders and creativity in artists, writers, and composers.

As a final note, the brain serotonin system may be involved in some nonclinical aspects of human behavior. Why do some people endlessly engage in rhythmic leg bouncing? What is rewarding about chewing gum? What underlies the therapeutic or reinforcing effects of breathing exercises and the twirling or dancing movements employed by various cults and religious groups? The reader can probably think of other behaviors that increase serotonin release in his or her brain.

Bibliography

Jacobs, B. L. 1991. Serotonin and behavior: Emphasis on motor control. *Journal of Clinical Psychiatry.* 52:17–23.

Jacobs, B. L., and E. C. Azmitia. Structure and function of the brain serotonin system. *Physiological Reviews* 72:165–229.

Jacobs, B. L., and C. A. Fornal. 1993. 5-HT and motor control: A hypothesis. *Trends in Neuroscience* 16:346–352.

Jacobs, B. L., P. J. Gannon, and E. C. Azmitia. 1984. Atlas of serotonergic cell bodies in the cat brainstem: An immunocytochemical analysis. *Brain Research Bulletin* 13:1–31.

Steinfels, G. F., J. Heym, R. E. Strecker, and B. L. Jacobs. 1983. Raphe unit activity in freely moving cats is altered by manipulations of central but not peripheral motor systems. *Brain Research* 279:77–84.

Trulson, M. E., B. L. Jacobs, and A. R. Morrison. 1981. Raphe unit activity during REM sleep in normal cats and in pontine lesioned cats displaying REM sleep without atonia. *Brain Research* 226:75–91.

Wilkinson, L. O., and B. L. Jacobs. 1988. Lack of response of serotonergic neurons in the dorsal raphe nucleus of freely moving cats to stressful stimuli. *Experimental Neurology* 101:445–457.

THOUGHT EVOKERS

- Explain the role serotonin plays in normal nerve cell activity.

- How is normal communication between neurons mediated? Explain.

- What was the first significant breakthrough regarding the behavior of serotonin neurons in the brain? Explain.

25 The evolution of aggression

*Humans are primates, and like our cousins
we use violence and cooperation in complex ways*

By William F. Allman

At Gombe National Park in Tanzania, a patrol of seven chimpanzees stealthily roved deeper and deeper into enemy territory. Suddenly, they heard the call of an infant chimpanzee nearby. Instantly rushing toward the sound, the chimps fell upon a hapless mother who was desperately trying to escape with her child. Over the next 15 minutes the chimps hit, bit and stamped on their victim and dragged her infant through the underbrush until the mother finally broke free. Bleeding profusely, she scooped up her wailing offspring and fled.

Lethal group-against-group violence was once thought to be uniquely human behavior. But primatologist Jane Goodall's recounting of "warfare" among neighboring groups of chimpanzees several years ago suggested for the first time that when it comes to aggression, the dividing line between humans and the rest of the animal kingdom is not clearly drawn. Goodall watched in horror as a band of chimpanzees she had been observing for years systematically hunted down individuals from a neighboring group and murdered them one by one, eventually wiping out the entire community.

Scientists have long used monkeys and apes as stand-ins for humans in testing food and drugs and researching basic human biology, but today they are turning more and more to primates for clues to human behavior as well. They have been especially interested in what primate aggression can say about human acts of violence. Their findings are providing a compelling counterargument to the age-old notion that human violence is either the consequence of innate biological drives or aberrant behavior caused solely by social pressures such as overcrowding and poverty. Instead, the new research suggests that aggressive behavior is a complex, inseparable mixture of both. What's more, aggression appears to be a vital component in maintaining cooperative relations among members of society.

> . . . aggression appears to be a vital component in maintaining cooperative relations among members of society.

Not everyone is enthusiastic about the new findings. Even though the seminal work of Charles Darwin is now a century and a half old, many people are still uncomfortable with the idea that humans share traits and ancestry with apes and monkeys, in spite of the fact that humans and apes have nearly identical DNA. Indeed, the U.S. government's top-ranked behavioral scientist recently resigned amid criticism of his controversial remarks suggesting that research on monkeys might yield clues to the problem of violence in inner cities—a suggestion that two prominent senators labeled "preposterous."

Clues from cousins. Despite such sentiments, primatologists have found that while studies of monkeys and apes may not provide specific prescriptions for solving human social problems such as gang violence, they can serve as an important window on certain human characteristics. "Although chimps are not simply little people in furry suits," observes Irwin Bernstein of the University of Georgia, "the basic building blocks of human behavior are present in their behavior."

One of the most surprising findings from recent animal research is that group violence and social cohesion are intimately linked. According to primate researchers Joseph Manson of the University of Michigan and Richard Wrangham of Harvard University, male chimpanzees who are best able to form alliances are also the most successful in competing for access to females and resources such as food and water, which are attractive to females. Cooperation, they say, appears to be a prerequisite for their successful aggression. A comparison of 42 foraging societies by Manson and Wrangham revealed the same dynamic at work: Males mount attacks—over females and vital resources—only when they have pieced together a coalition that outnumbers the outsiders, so there is little risk to the attackers.

Manson and Wrangham are careful to point out that their work does not mean that chimp and human aggression are identical; humans are far more sophisticated in their cooperation, technology, and the kinds of resources they fight over. But the evidence that apes weigh risks and form coalitions suggests that all primate aggression, including human aggression, is more complex than a simple bubbling up of some atavistic animal instinct or a mere reaction to stress. It is instead, the research suggests, an ancient evolutionary strategy more closely tied to coalition building and harmony than to murderousness and wanton violence. Nor is aggression a trait found just among male primates: Studies show that in many ape and monkey societies, the females, not males, band together to fight aggressively over resources.

In contrast to the causes of group-against-group conflicts, aggression within a single primate community appears to serve as a tool for maintaining the group's social stability. Research by the University of Georgia's Bernstein and Tom Gordon of Emory University's Yerkes Primate Center demonstrates that in rhesus monkeys, for instance, most aggressive encounters take place when a newcomer attempts to join an established group. Rhesus monkeys live in a strict peeking order, and newcomers must nudge and jostle their way into the hierarchy. Once the monkey is established in the group, however, the level of aggression drops dramatically, as each individual generally defers to those higher on the social ladder. Gordon points out that deferential encounters occur three times more often than do encounters involving overt acts of aggression.

A monkey's ability to establish itself in the social hierarchy depends far less on its physical strength, aggressiveness, or fighting ability than on its ability to make friends. Older males, for instance, often maintain their social position by craftily allying themselves with other influential monkeys in the group. "During a power struggle, males will try to get support of females by playing with their infants," says Yerkes primatologist Frans de Waal. "Presidential candidates do the same thing."

Monkeys will also band together to topple a higher-up. In one experiment, a high-ranking female was removed from her allies and placed among a group of low-ranking females, who quickly formed an alliance against her and reversed the hierarchy. Again, it was cooperation rather than untempered aggression that guaranteed success.

Ultimately, social cohesion among primates requires that allies play straight with one another, and aggres-

sion appears sometimes to have a key role in discouraging cheating. In one instance, a chimp that had formerly come to the aid of another chimp during a fight came under attack himself. When he extended his hand toward his partner in an apparent plea for help and was ignored, he turned and attacked his betrayer. Another study by de Waal found that when a bundle of food was placed among a group of chimpanzees, the greatest amount of aggression was used not to fight for a fair share but to exclude those chimps who had refused to share their food on a previous occasion.

Mothers, too, use aggressive punishment to teach social lessons that will ultimately smooth their offspring's passage in the group. Indeed, experiments have shown that rhesus monkeys who are taken from their mothers as infants and raised in isolation become social misfits who are unable to make allies when they are later put into a group.

Primatologist Stephen Soumi of the National Institute of Mental Health has found that the genes of the father can also influence how a monkey gets along in a group. In a typical group of rhesus monkeys, there is usually a small percentage of males who do not get along with anyone. Like the monkeys who are raised in isolation, these monkeys disregard the group's social structure, frequently engage in aggressive acts without provocation, and sometimes are expelled from the group. These monkeys are often the offspring of fathers who display similar traits, says Soumi.

Brain chemistry. Analyzing the spinal fluid of these antisocial monkeys, Soumi has found that they have low levels of a particular byproduct of a brain hormone called serotonin. In contrast, monkeys higher up in the

social hierarchy have higher levels of the same brain chemical. Intriguingly, low levels of the same brain hormone have been discovered in Marines discharged from the corps for excessive violence, and a study of criminals in Finland who committed acts of wanton violence also found they had low levels of this brain chemical.

There is no evidence yet that low levels of serotonin directly cause violent behavior, Soumi emphasizes. While hormones can certainly produce behavioral changes, he says, social circumstances can also cause changes in brain chemistry. Poor nurturing, for instance, is known to lower a monkey's serotonin level, while good nurturing can raise it. "The whole debate over nature versus nurture is meaningless," says Soumi. "It's not either/or, it's *and*."

Ultimately, the same could be said for the supposed distinction between aggression and peaceful coexistence. Rather than being polar opposites, these behaviors in fact are woven together in the complex interactions through which humans—and many other primates—maintain balance in their social relations. Indeed, as much as nonhuman primates might reveal about aggression, they tell far more about how close-living individuals can resolve their inevitable conflicts of interest peacefully, says Yerkes's de Waal, who has observed many instances of chimpanzees reconciling after an aggressive encounter. Like other primates, humans must continually work to control harmful aggression within their societies. Understanding the biological and cultural roles that aggression plays in human life will ultimately be more constructive than moralistic efforts to ban this fundamental primate behavior.

THOUGHT EVOKERS

- Why are scientists using research on primates in order to better understand human aggression?

- How are "group violence" and "social cohesion" linked to animal aggression?

DATE DUE

Demco, Inc. 38-293